Question Bank on General Agriculture and Engineering

Question Bank on General Agriculture and Engineering

(Special focus on IBPS-AFO Exam)

Also Usefull for
JRF, SRF, NET, ARS, IFS, FCI, AIC, IIFCO, NSCI, ADO, D.H.O, AEO and all Similar Competitive Agricultural Related Entrance Exams.

Gurdeep Lal
Shivendu Pratap Singh Solanki
Shivkumar G. Telkar
Yogesh Shivajirao Nilvarn
Rashmi Sahu
Shashikant Sharma

SSPH

SATISH SERIAL PUBLISHING HOUSE
403, Express Tower, Commercial Complex
Azadpur, Delhi-110033 (India)
Phone : 011-27672852, Fax : 91-11-27672046
e-mail : info@satishserial.com, hkjain1975@yahoo.com
Website : www.satishserial.com

Published by :

SATISH SERIAL PUBLISHING HOUSE
403, Express Tower, Commercial Complex, Azadpur, Delhi-110033 (INDIA)
Phone : 011-27672852 Fax : 91-11-27672046
E-mail : info@satishserial.com, hkjain1975@yahoo.com

ISBN 978-93-86200-55-6

Composed, Designed & Printed in India

About the Authors

Gurdeep Lal, working as an Agricultural Field officer, Bank of Baroda, born at village Kazibans, Saharanpur District, Utter Pradesh. He received his Post Graduation degree (2015-17) in Farm Machinery and Power Engineering from Department of Agriculture and food Engineering IIT Kharagpur West Bengal and Under Graduate degree (2011-15) in Agricultural Engineering from College of Agricultural Engineering and Technology, Vasantrao Naik Marathwada Agricultural University, Parbhani (Maharashtra). He was awarded with ICAR-UG-NTS for his B.Tech. His schooling done in Jawahar Nawodaya Vidyalaya Saharanpur (2004-2011). He has qualified the GATE AIR 132, ICAR-JRF AIR 65, in year 2015. He is the founder and tutor of YouTube Channel Advance Agricultural Classes, and mentor of educational website http://www.advanceagriclasses.com

Shivendu Pratap Singh Solanki Ph.D Scholar, Punjab Agricultural University, Ludhiana, born at village Lamkana, Katni District, Madhya Pradesh. He received his under graduate degree in Horticulture (2011-15) from College of Horticulture, Vasantrao Naik Marathwada Agricultural University, Parbhani (Maharashatra) and awarded with University Gold Medal and Best Student award for excellent academic performance. He was awarded the ICAR-JRF for his M.Sc. (Horticulture) in Fruit Science at Dr. Yaswant Singh Parmar University of Horticulture and Forestry, Nauni, Solan (Himachal Pradesh) in 2015-17. He has qualifiedthe ICAR-ASRB-NET in Fruit Science and also qualified ICAR-PGS-SRF examination with AIR 2nd rank in year 2017. He is advisory board member and reviewer of International Journal of Agricultural Sciences and Assistant editor of International Journal of medicinal plant studies. He is also a life member of various reputed scientific societies . He has published 7 popular articles and 5 research article in national and international journals and awarded with Excellence AIASA Cabinet Member Award-2017. He is mentor of popular educational website Horticulture Gyan (https://horticulturegyan.wordpress.com/).

Shivkumar G. Telkar, born at Nandura, Buldhana District, Maharashtra. He received his under graduate degree in Agriculture (2011-15) from Shri Shivaji Agriculture college, Amravati, Dr. Punjabrao Deshmukh Krishi Vidyapeeth, Akola (Maharashatra) and he was awarded the ICAR-PG-NTS for his M.Sc. (Agriculture) in Agronomy at college of post graduate studies, CAU, Imphal in 2015-17. He has qualified the ICAR-ASRB-NET in Agronomy and also qualified ICAR-PGS-SRF examination. He has published 12 popular articles in reputed international news-letters.

Yogesh Shivajirao Nilvarn, born at village Mangarul, Parbhani District, Maharashtra. He received his under graduate degree in Horticulture (2011-15) from College of Horticulture, Vasantrao Naik Marathwada Agricultural University, Parbhani (Maharashatra). He was awarded the ICAR-JRF for his M.Sc. (Horticulture) in Floriculture and Landscaping at The Mahatma Phule Krishi Vidyapeeth, *Rahuri* (Maharashtra) in 2015-17.

Rashmi Sahu, working as an Assistant Professor in SVKR Collage, born at District Jabalpur, Madhya Pradesh. She received her Post Graduation degree (2015-17) in Farm Machinery and Power Engineering from Department of Agriculture and food Engineering IIT Kharagpur West Bengal and Under Graduate degree (2011-15) in Agricultural Engineering from College of Agricultural Engineering, Jawaharlal Nehru Krishi Vishwa Vidyalaya, Jabalpur (Madhya Pradesh). She has qualified the GATE AIR 23, ICAR-JRF AIR 70, in year 2015.

Shashikant Sharma Ph.D Scholar, Junagadh Agricultural University, Junagadh (Gujarat), born at village Shivpur, Bhojpur District, Bihar. He received his under graduate degree in Agriculture (2011-15) from AC&RI, Killikulam, Tamil Nadu Agricultural University, Coimbatore (Tamil Nadu). He was awarded the ICAR-PG-NTS for his M.Sc. (Biotechnology) at Dr. Yashwant Singh Parmar University of Horticulture and Forestry, Nauni, Solan (Himachal Pradesh) in 2015-17. He has qualified the ICAR-PGS-SRF examination with AIR 11th rank in year 2017. He is also a life member of various reputed scientific societies. He has published 2 popular articles and research article in national and international journals.

Preface

This is the epoch of smart and tough competition. Competition is exploding in all fronts and in the modern times everyone wants to get succeed in competition. Agriculture is the vast subject for preparation of competitive examination. To get the success in agricultural competition becoming selective for study is the astute move and the only conduit for success. The ICAR, New Delhi is conducting various examinations like ICAR-JRF, SRF, NET examination etc. In addition to that Government of India and State Government conduct various exams like IFS, FCI, AIC, IIFCO, NSCI, ADO, D.H.O, AEO and all Similar Competitive Agricultural Related Exams and IBPS-AFO exam every year conducted by Government of India. All these examinations are mostly based on Multiple type questions and students always look for study material that is ready to read and directly useful for examination. Keeping in view, it felt to bring out a book which could provide the basic and innovative knowledge of various fields of agriculture and allied sciences.

The present book "**Question Bank on General Agriculture and Engineering**" has been prepared in most simple and appropriate manner by taking the view of students, teachers and scientists. We hope that the book will fulfill the need of students appearing in agricultural based competitive examinations.

The help and prop up extended by our family members, friends, student, teachers and scientists during the compilation of this book is highly appreciable.

We humbly welcome the valuable suggestions from readers for further improvement of this book.

Authors

Contents

Chapter 1

Agricultural Economics

1. **National markets are found for**
   a) Perishable goods b) Food grain
   c) Durable goods d) Pulse crops

2. **Regional markets are found for**
   a) Perishable goods b) Food grain
   c) Durable goods d) Pulse crops

3. **Short period markets are found for**
   a) Perishable goods b) Food grain
   c) Durable goods d) Pulse crops

4. **Which market is permanent in nature?**
   a) Short period market b) Secular market
   c) Long period market d) None of the above

5. **Food grain markets, vegetable markets, wool markets are the example of**
   a) Regional market b) Specialized market
   c) General market d) World market

6. **A market in which there are more than two but still a few sellers of a commodity is termed as**
   a) An oligopsony market b) Monopolistic market
   c) Monopoly market d) An oligopoly market

7. **Indian farmers operate in __________ market when purchasing electricity for irrigation.**
   a) Oligopoly b) Monopoly
   c) Duopoly d) Monospony

8. **Transportation adds __________ utility**
   a) Place b) Time
   c) Possession d) Form

9. **The marketing function of buying and selling adds __________ utility**
   a) Place b) Time
   c) Possession d) Form

10. **__________ is the first function performed in the marketing of agricultural commodities.**
    a) Financing
    b) Packing
    c) Grading and standardization
    d) Transportation

11. **__________ who take title of the product with a view to making a profit.**
    a) Agent middlemen
    b) Facilitative middlemen
    c) Speculative middlemen
    d) Merchant middlemen

12. **The FCI was established in the year**
    a) 1955 b) 1965
    c) 19 d) 1935

13. **A market in which there are more than two but still a few buyers of a commodity is termed as**
    a) An oligopsony market b) Monopolistic market
    c) Monopoly market d) An oligopoly market

14. **Regulated market committee consisting of ________ members.**
    a) 17 b) 27
    c) 19 d) 7

15. **GATT and WTO established in the year**
    a) 1937 & 1995 b) 1947 & 1995
    c) 1937 & 1985 c) 1947 & 1985

16. **The Warehousing Corporations Act came into operation on**
    a) 18th March, 1962 b) 18th March, 1952
    c) 18th May, 1962 d) 18th March, 1972

17. **Commercial banks advance up to ______ percent of the value of the produce stored in the warehouse.**
    a) 65 b) 75
    c) 80 d) 70

18. **Regularly occurring upswings and downswings in prices are termed as ______fluctuations in prices.**
    a) Cyclical b) Annual
    c) Seasonal d) Short-term

19. **_____arise due to changes in Government policies, programmes, changes in taxes and tariffs**
    a) Marketing risk b) Price risk
    c) Trade risk d) Institutional risks

20. **Central Agmark Lab is located at**
    a) Mumbai b) Nagpur
    c) Delhi d) Faridabad

21. **The Agricultural Produce (Grading and Marketing) Act was passed in the year**
    a) 1927 b) 1957
    c) 1937 d) 1917

22. **The Prevention of Food Adulteration Act was passed in the year**
    a) 1954 b) 1944
    c) 1934 d) 1964

23. **Vegetable Oil Products (Standard of Quality) Order was passed in the year**

a) 1965 b) 1945
c) 1955 d) 1975

24. **The area of operation of society is restricted to one village is called**

a) PACS b) Apex bank
c) Central bank d) None

25. **The statutory minimum price is announce for**

a) Potato b) Wheat
c) Sugar beet d) Sugarcane

26. **The marketed surplus is _______ to the marketable surplus for perishable commodities.**

a) Equal b) Greater than
c) Less than d) Greater than or less than

27. **DMI headquarter is located at**

a) Hyderabad b) Faisalabad
c) Faridabad d) Mumbai

28. **Marketable surplus is given by**

a) MS = P + C b) MS = P – C
c) MS = C – P d) MS = P – C / 2

29. **WTO came in to effect from**

a) April 1995 b) March 1995
c) December 1995 d) January 1995

30. **Risks arise due to changes in Government policies are termed as a**

a) Institutional risk b) Physical risk
c) Price risk d) None

31. **The Warehousing Corporations Act came into operation on**
    a) 18th April, 1962 b) 18th March, 1962
    c) 18th January, 1962 d) 18th February, 1962

32. **The first State Warehousing Corporations was set up in**
    a) Gujarat, 1956 b) Punjab, 1956
    c) Bihar, 1956 d) Rajasthan, 1956

33. **International Organization for Standardization (ISO) came into existence on**
    a) 25th February, 1947 b) 25th July, 1947
    c) 25th January, 1947 d) 25th June, 1947

34 **The Codex Alimentations Commission (CAC) was established in**
    a) 1973 b) 1963
    c) 1953 d) 1983

35. _________ **is not a function of marketing.**
    a) Processing b) Financing
    c) Harvesting d) Assembling

36. _________ **is considered as market lifeblood)**
    a) Buyers b) Market information
    c) Sellers d) Market organization

37. **What is "support price" for an agriculture commodity?**
    a) Subsidy paid by the government over
    b) The floor price below which it can the price already available in the marketnot be sold
    c) The minimum price at which the Government is prepared to buy it
    d) Money paid to agriculturists for case of draught damaging their crops

38. **Which one of the following is a competitive market?**
    a) Perfect market b) Primary market
    c) Wholesale market d) Capital market

39. **Regulated market ensures**
    a) Procurement price b) Support price
    c) Remunerative price d) Fair price

40. **The Agricultural Price Commission was set up in the year**
    a) 1955 b) 1965
    c) 1975 d) 1970

41. **Codex Alimentations Commission established on**
    a) 1963 b) 1947
    c) 1946 d) 1935

42. **The Fruit Products Order, (1956) order lays down mandatory standards for**
    a) Fresh fruits b) Canned fruits
    c) Processed fruits d) None

43. **The Indian Seeds Act was passed in the year**
    a) 1976 b) 1956
    c) 1966 d) 1986

44. **NAFED was established in the year**
    a) November, 1958 b) October, 1958
    c) August, 1958 d) December, 1958

45. **In our state the structure of co-operative marketing is**
    a) One-tier system b) Two-tier system
    c) Three-tier system d) Four-tier system

46. **Directorate of Marketing and Inspection established on**
    a) 1965 b) 1935
    c) 1945 d) 1995

47. **Food Corporation of India established on**
    a) 1965 b) 1935
    c) 1946 d) 1990

48. **International Organization for Standardization established on**

a) 1947 b) 1945

c) 1946 d) 1990

49. **Processing adds following utility**

a) Form b) Place

c) Time d) None of above

Answers

1 c	2 b	3 a	4 b	5 b	6 d	7 b	8 a	9 c	10 b
11 c	12 b	13 a	14 a	15 b	16 a	17 b	18 a	19 d	20 b
21 c	22 a	23 d	24 a	25 d	26 a	27 c	28 b	29 d	30 a
31 b	32 c	33 a	34 b	35 c	36 b	37 c	38 a	39 d	40 b
41 a	42 c	43 c	44 b	45 c	46 b	47 a	48 a	49 a	

❑❑❑

Chapter 2

Agricultural Meteorology

1. **Solar radiation produces by which process?**
   a) Adiabatic process b) Nuclear Fusion
   c) Photoelectric effect c) Chaos process

2. **Solar energy travels to space in discrete packets of energy known as?**
   a) Photon (Electromagnetic wave)
   b) Boson
   c) Fermion
   d) Meson

3. **What is solar constant on earth?**
   a) 100 Langley b) 1.94 Langley
   c) 22 Langley d) 2.987 Langley

4. **A form of energy that is emitted by all objects having a temperature above absolute zero known as?**
   a) Convection b) Radiation
   c) Conduction d) None of these

5. **What is the wavelength of visible solar radiation?**
   a) 400-700 nm b) 1000 nm
   c) 100-150 nm d) 2.13-5.0 nm

6. **Which law states that a good absorber of radiation is also a good emitter under similar condition?**
   a) Planck's law b) Kirchoff's law
   c) Stefens's law d) Weins's displacement law

7. **Which one of these wavelengths has maximum energy?**
   a) 0-300 nm b) 400-430 nm
   c) 1000-á nm d) 189000 nm

8. **Solar radiation measured by?**
   a) Pyranometer b) Thermometer
   c) Hygrometer d) Barometer

9. **A certain part of energy received from the sun, is reflected back to space by the earth known as?**
   a) Transmission b) Absorption
   c) Albedo d) Conduction

10. **By which process green plants and certain other organisms use the energy of light to convert carbon dioxide and water into the simple sugar glucose?**
   a) Transpiration b) Detoxification
   c) Photosynthesis d) Respiration

11. **According to which law the energy flux density of radiation from a body is a function of fourth power of its absolute temperature?**
   a) Planck's law b) Kirchoff's law
   c) Stefens's law d) Weins's displacement law

12. **The instrument able to record almost all meteorological data by desired interval at any time and any place**
   a) Automatic weather station
   b) Stevenson screen
   c) Ceilometer
   d) Evaporimeter

13. **Which instrument record the wind direction**
   a) Anemometer b) Wind Vane
   c) Thermograph d) Barograph

14. **Which instrument used for measure wind speed**
   a) Barometer b) Hygrograph
   c) Anemometer d) Radiosonde

15. **Instrument used for record atmospheric pressure**
    a) Luxmeter b) Barometer
    c) Photometer d) Quantum Sensor

16. **Relative humidity measured by**
    a) Solarimeter b) Hygrometer
    c) Pyranometer d) Pyrgeometer

17. **Evaporation measured by**
    a) Pyrgeometer b) Evaporimeter
    c) Albedometer d) Radiosonde

18. **Continuous temperature record by which instrument**
    a) Thermograph b) Barograph
    c) Hygrograph d) Luxmeter

19. **Which Instrument record temperature without contact the object**
    a) Infrared thermometer b) Evaporimeter
    c) Albedometer d) Radiosonde

20. **Combination of Dry bulb and Wet bulb thermometer used for**
    a) Net radiation b) Relative Humidity
    c) Cloud d) Gas Composition

21. **Gaseous envelop surrounding the earth known as.**
    a) Atmosphere b) Biosphere
    c) Hydrosphere d) Grid

22. **In physics, an idealized object that absorbs all electromagnetic radiation falling on it known as?**
    a) Black body b) Thermodynamics
    c) Isenthalpic process d) Entropy

23. **The circulation of the atmosphere occurs due to?**
    a) Kinetic energy b) Inertial frame of reference.
    c) Gottfried Leibniz d) Thermal differences

24. **What is the height of troposphere?**

a) 16-30 km b) 30-40 km
c) 0-16 km d) 40-90 km

25. **Which phenomena is responsible for maintaining earth's temperature for life?**

a) Detoxification b) Greenhouse effect
c) Procurement d) Assortment

26. **What is the dry adiabatic lapse rate in troposphere?**

a) 6.5°C/km b) 10.0°C/km
c) 15.2°C/km d) 25.0°C/km

27. **In which part of atmosphere the ozone layer located?**

a) Troposphere b) Mesosphere
c) Thermosphere d) Stratosphere

28. **In the atmosphere, which of the following gases account for about 99.0 % percent by volume?**

a) Nitrogen, Oxygen, Carbon dioxide
b) Hydrogen, Methane, Helium
c) Argon, Neon, Krypton
d) Methane, Helium, Xenon

29. **Plants produce oxygen through**

a) Respiration b) Photosynthesis
c) Transportation d) Excretion

30. Which type of gases is highly detrimental for ozone layer?

a) Halogenated gases b) Liquefied gases
c) Incendiary gases d) Ruling

31. The percentage of water vapour in atmosphere by volume?

a) 0-4% b) 0-10 %
c) 1-30% d) 0-.5%

32. **What is the reason, the incoming solar radiation in the form of short wave radiation but after striking the earth surface it will be converted into long wave radiation?**

a) Surface temperature b) Humidity
c) Cloud d) Water

33. **What is the height of atmosphere?**

a) 0-500 km b) 0-1000 km
c) 0-4000 km d) 0-1000000 km

34. **In which part of atmosphere all types of weather phenomena occur?**

a) Mesosphere b) Stratosphere
c) Troposphere d) Exosphere

35. **An aggregation of minute drops of water suspended in the air at higher altitude termed as.**

a) Clouds b) Fog
c) Frost d) Haze

36. **Unit Okta used to record which weather phenomena**

a) Clouds b) Mist
c) Smog d) Miasma

37. **Imaginary line that represents the equal rainfall termed as.**

a) Isohyet b) Isochrome
c) Isophane d) Isotech

38. **World meteorological organization (WMO) classified the cloud in**

a) 5 parts b) 8 parts
c) 3 parts d) 10 parts

39. **India receive maximum rainfall by**

a) South West Monsoon b) North East Monsoon
b) Cyclones d) Hurricane

40. **Which cloud known as rainy cloud**
   a) Altocumulus b) Cumulonimbus
   c) Cirrostratus d) Altostratus

41. **In which part of atmosphere rainy clouds are formed**
   a) Troposphere b) Mesosphere
   c) Thermosphere d) Stratosphere

42. **The process in which the water vapour is converted into its liquid termed as**
   a) Condensation b) Concentration
   c) Compression d) Reduction

43. **Name of science's branch in which, the collection and interpretation of information about a target without being in physical contact with it?**
   a) Remote Sensing b) Quantum Mechanics
   c) Astrophysics d) Nuclear Physics

44. **Satellite which move around to North Pole to South Pole known as?**
   a) Polar Satellite b) Navigator Satellite
   c) Scientific Satellite d) Weather Satellite

45. **What is the name of world first artificial satellite?**
   a) Sputnik 1 (Oct 4, 1957) b) TIROS
   c) METSAT d) GOES

46. **The 100 meter length of represent by 10 centimeter on the map called as?**
   a) Trim b) Scaling down
   c) Dapper d) Magnitude

47. **At which height the satellites move in space?**
   a) 36000 km b) 50000 km
   c) 45000 km d) 60000 km

48. **Solar radiation produces by which process?**
   a) Adiabatic process b) Nuclear Fusion
   c) Photoelectric effect d) Chaos process

49. **Solar energy travels to space in discrete packets of energy known as?**
   a) Photon (Electromagnetic wave)
   b) Boson
   c) Fermion
   d) Meson

50. **Which green house gas linked with rice crop?**
   a) Neon b) Xenon
   c) Argon d) Methane

51. **The latent heat of condensation is released in air under the atmospheric condition of _____.**
   a) Dry adiabatic lapse rate
   b) Wet adiabatic lapse rate
   c) Inversion
   d) Isothermal lapse rate

52. **In stratosphere temperature increases due to absorption of _____.**
   a) Ultraviolet radiation b) V visible radiation
   c) Infrared radiation d) Microwave radiation

53. **Surface has the lowest albedo in visible radiation ________.**
   a) Soil b) Vegetation
   c) Water d) Ice

54. **Secondary circulation of atmosphere over earth's surface is called _____.**
   a) Esterlies
   b) Westerlies

c) Cyclones & anticyclones

d) Land & sea Breezes.

55. **Line joining the places having equal atmosperic pressure is known as _____**

a) Isotherm b) Isobars

c) Isohytes d) Isonymph

56. **Instantaneous physical state of atmosphere is called______.**

a) Climate b) Microclimate

c) Ecoclimate d) Weather

57. **Albedo values of agricultural field ranges between _________.**

a) 0.18-0.27 b) 0.29-0.35

c) 0.40-0.50 d) 0.51-0.65

58. **Clouds types which gives the heavy and continuous precipitation ----**

a) Cumulus b) Nimbus

c) Cumulonimbus d) Stratocumulus

59. **Summation of mean temperature above threshold temperature is called**

a) GDD b) AET

c) PET d) TUE

60. **Optimum temperature for wheat crop is between ____.**

a) 10- 150°C b) 15- 20 °C

c) 21-26 °C d) 26- 30 °C.

61. **Distribution of radiation within plant canopy is estimated with the help of**

a) Beer's law b) Kirchoff's law

c) Stefan boltzman law d) Plank's law.

62. **Change of phase from solid to vapour or vice versa is known as ____.**

a) Evaporation b) Condensation
c) Precipitation d) Sublimation

63. **The average sea level pressure is equal to ______.**

a) 1103 mb b) 1013 mb
c) 1030 mb d) 1003 mb

64. **Heat transfer in soil takes place mainly through the process of ____.**

a) Convection b) Advection
c) Radiation d) Conduction

65. **Heat flux which remains dominant over a dry barren field is _____.**

a) Sensible b) Latent
c) Ground d) Photosynthetics

Answers

1 b	2 a	3 b	4 b	5 a	6 b	7 a	8 a	9 c	10 c
11 c	12 a	13 b	14 c	15 b	16 b	17 b	18 a	19 a	20 b
21 a	22 a	23 d	24 c	25 b	26	27 d	28 a	29 b	30 a
31 a	32 a	33 a	34 c	35 a	36 a	37 a	38 d	39 a	40 b
41 a	42 a	43 a	44 a	45 a	46 b	47 a	48 b	49 a	50 d
51 b	52 a	53 c	54 c	55 b	56 d	57 a	58 a	59 a	60 b
61 a	62 d	63 b	64 d	65 a					

❑❑❑

Chapter 3

Agril Engineering

1. **Animals if raised by the farmer himself can be a very cheap source of**

   a) Transport power b) Cart power

   c) Farm power d) Bullock power

2. **Acceptability of the tractor and power tiller is not very common because of high initial**

   a) Price b) Coverage

   c) Investment d) Power requirement

3. **The average capacity of a wind mill would be about**

   a) 2.0 hp b) 1.5 hp

   c) 1.0 hp d) 0.5 hp

4. **Wind mill having 3.6md iameterw heelmounted on 12m twerisable to produce 0.1 to 0.9 hp with a wind velocity varying from**

   a) 2.4-27 kmph b) 1.4–37 kmph

   c) 2.4-37 kmph d) 3.4–47 kmph

5. **Intake valves wear more than exhaust valves is probably due to performance by the**

   a) Air nozzles b) Air cleaner

   c) Top dead center d) Bottom dead center

6. **Intake valves wear more than**

   a) Valve seats b) Valve grinding plates

   c) Exhaust valves d) Outlet valves

7. **Tractor engine compression pressure goes down considerably after constant operation of about**
   a) 1000 hours b) 1500 hours
   c) 2000 hours d) 2500 hours

8. **Clearance between the rocker arm and valve stem is known as**
   a) Rocker clearance b) Stem clearance
   c) Tappet valve clearance d) Liner clearance

9. **The valve tappet clearance is adjusted when both the valves arc in**
   a) Open position b) Close position
   c) Shut position d) Stem open position

10. **A typical valve of tappet clearance is given as**
   a) 0 .28 mm b) 0.38 mm
   c) 0 .48 mm d) 0 58 mm

11. **The valve face is generally made of heat resisting material like**
   a) Carbon steel b) Stainless steel
   c) High carbon steel d) Special alloys

12. **The top of the piston is called the crown and the lower section the**
   a) Bottom cover b) Bottom space
   c) Skirt d) Brush

13. **Piston and piston rings seal the cylinder and transmit the gas pressure to the**
   a) Cam shaft b) Crank shaft
   c) Crank case d) Connecting rod

14. **Close contact between the dry liner and the cylinder is absolutely necessary to reduce the resistance to**
   a) Wear b) Tear
   c) Oil flow d) Heat flow

15. **Weight of two stroke engine per hose power is**

a) High b) Low
c) Medium d) Optimum

16. **Weight of four stroke engine per horse power is**

a) Low b) High
c) Optimum d) Medium

17. **Type of two stroke engines are available only as**

a) All speed b) Low speed
c) Medium sped d) High speed

18. **Type of four stroke engine available through manufacturers are**

a) Only high speed b) Only low speed
c) Medium spend d) All speeds

19. **Four stroke cycle engines can be operated in the direction of**

a) Only one b) Only opposite
c) Clockwise d) Counterclockwise

20. **Flywheel of four stroke cycle engine is comparatively**

a) Larger b) Smaller
c) Medium d) Optimum

21. **Flywheel of two stroke cycle engine is comparatively**

a) Smaller b) Large
c) Medium d) Optimum

22. **Thermal efficiency of two stroke cycle engine is comparatively**

a) High b) Low
c) Minimum d) Maximum

23. **Thermal efficiency of four stroke cycle engine is comparatively**

a) Higher b) Lower
c) Minimum d) Maximum

24. **Operating cost of diesel engine is comparatively**
   a) High b) Very high
   c) Low d) Optimum

25. **Operating cost of carburetor engine is comparatively**
   a) Low b) High
   c) Very high d) Maximum

26. **Initial cost of diesel engine is comparatively**
   a) Low b) High
   c) Minimum d) Maximum

27. **Diesel engine weight per horse power is**
   a) Low b) High
   c) Maximum d) Minimum

28. **Weight of carburetor engine per horse power is comparatively**
   a) High b) Low
   c) Very low d) Very high

29. **After being heavily loaded diesel engine develops more**
   a) Draft b) Force
   c) Pull hour d) Torque

30. **The top of the piston is called the crown and the lower section the**
   a) Bottom cover b) Bottom space
   c) Skirt d) Brush

31. **Piston and piston rings seal the cylinder and transmit the gas pressure to the**
   a) Cam shaft b) Crank shaft
   c) Crank case d) Connecting rod

32. **The ratio of the horse power output of an engine to the fuel horse power is known as**
   a) Mechanical efficiency b) Brake thermal efficiency
   c) Volumetric efficiency d) Fuel consumption

33. **The quantity of fuel consumed by an oil engine on the basis of 1 horse power hour is known as**

a) Specific heat

b) Thermal equilibrium

c) Specific fuel consumption

d) Air-fuel mixture

34. **The ratio of the brake horse power to the indicated horse power is called**

a) Brake thermal efficiency

b) Volumetric efficiency

c) Denial efficienc

d) Mechanical efficiency

35. **The horse power available at the crank shaft is called**

a) Crank horse power b) Brake horse power

c) Horse power d) Indicated horse power

36. **The total horse power developed by allthecy linders and needed piston, disregarding friction and losses within the engine is said to**

a) Horse power b) Indicated horse power

c) Frictional horse power d) Brake horse power

37. **The average net pressure in kg/m^2 on the piston during the power stroke only is called**

a) Brake mean effective pressure

b) Indicated mean effective pressure

c) Engine pressure

d) Cylinder pressure

38. **Tractors are mostly equipped with**

a) Single cylinder engines b) Two cylinder engines

c) Three cylinder engines d) Multiple cylinder engines

39. **Tractor engine compression pressure generally goes down considerably after constant operation of about**

a) 1000 hours b) 1500 hours

c) 2000 hours d) 2500 hours

40. **Two stroke engine have very limited use on**

a) Dairy farm b) Cattle farm

c) General farms d) Specific farms

41. **Multiple cylinder, high speed, vertical engines are mostly used as automotive engines on**

a) Farm truck b) Feed grinding machines

c) Farm tractors d) Farm combines

42. **Some of the small size garden tractors are equipped with**

a) Single cylinder engines b) Four stroke engines

c) Two stroke engines d) Multiple cylinder engines

43. **The thermostat opens widely at about**

a) 75°C b) 78°C

c) 80°C d) 82°C

44. **The cooling system of an internal combustion petrol engine maintains the engine temperature in the range of**

a) 61 to 70°C b) 71 to 80°C

c) 74 to 78°C d) 78 to 82°C

45. **For heavier fuels the cooling system of an internal combustion maintains the engine temperature at**

a) 78 to 88°C b) 8O to 90°C

c) 88 to 90°C d) 90 to 92°C

46. **After release of press button diesel engine must start within**

a) 10 sees b) 12 secs.

c) 15 secs d) 17 secs

47. **If the engine does not start on the first attempt, subsequent attempt should be made at an interval of about**

a) One minute b) Two minutes
c) Three minutes d) Four minutes

48. **Magneto points need to be regapped alter operation of about**

a) 100 to 200 hours b) 150 to 300 hours
c) 200 to 300 hours d) 250 to 500 hours

49. **Using the pump for transporting oil in the system pressure develop in the lubricating system is about**

a) 1.2 kg/sq.cm. b) 1.5 kg/sq.cm.
c) 2.5 kg/sq.cm. d) 3 kg/sq.cm.

50. **Normally the thermostat begins to open at about 74°C and opens widely at about**

a) 75°C b) 78°C
c) 80°C d) 82°C

51. **A completely discharged battery will have electrotype of specific gravity of**

a) 1.150 or less b) 1.050 or less
c) 1.175 or less d) 1 .1 8 5 or less

52. **Where the intensity of dust is not severe generally pre-cleaners a placed above the tractor hood about**

a) 35 cm b) 40 cm
c) 45 cm d) 50 cm

53. **The minimum torque is produced at**

a) 1200 rpm b) 1300 rpm
c) 1400 rpm d) 1500 rpm

54. **Most of the time of the engine is required to be operated between**

a) 40 to 70 percent of full load
b) 50 to 70 percent of full load

c) 55 to 75 percent of full load
d) 60 to 90 percent of full load

55. **A tractor engine is operated at full load approximately**
a) 20-25% of the operating times
b) 25-30% of the operating time
c) 30-35% of the operating time
d) 35-40% of operating time

56. **The boiling temperature of the water in the radiator is raised to about**
a) 100°C b) 105°C
c) 110° C d) 115° C

57. **Due to constant use both inlet and exhaust valves and their seats subjected to**
a) Stress wear b) Strain wear
c) Mechanical wear d) Tensile wear

58. **For he avierfue ls the cooling system of an internal combustion engine, the temperature is maintained at**
a) 78to 80°C b) 80to 84 °C
c) 85to 88°C d) 88to 90°C

59. **In cranking the engine normally motor takes current about**
a) 300 amperes b) 400amperes
b) 5 0 0 amperes d) 600amperes

60. **A completely discharged battery will have electrotype of specific gravity of**
a) 1.150 or less b) 1.050 or less
c) 1.175 or less d) 1.185 or less

61. **When the engine is operated under average working conditions oil crank case us changed after the operation of about**
a) 100hours b) 110 hours
c) 115 hours d) 120 hours

62. **The maximum V belt sag when applying finger pressure is**
    a) 10 mm b) 12 mm
    c) 15 mm d) 17 mm

63. **The solution of radiator should be drained out when the engine has run for**
    a) 10 to 15 minutes b) 15to 20 minutes
    c) 20 to 25 minutes d) 25 to 30 minutes

64. **To make a solution for filling radiator one bag of washing soda and 0.5 kg of kerosene oil should be added in water amounting**
    a) 6 kg b) 8 k8
    c) 1 0 kg d) 12 kg

65. **The radiator should he filled with the solution and allowed to remain therefore**
    a) 4 to6 hours b) 6 to 6 hours
    c) 8 t o 10 hours d) 10 to 12 hours

66. **The starting motor used on a powerine tractor may be of**
    a) 6 or 12 volt type b) 12 or 24 volt type
    c) 16 or 24 volt type d) 6 or 24 volt type

67. **Diesel tractors are equipped with a more rugged starting motor of**
    a) 6 or 12 volts c) 12 or 24 volts
    c) 6 or 24 volts d) 12 or 36 volts

68. **Voltage can be stepped up from 6 to approximately**
    a) 10,000 volts b) 15.00 0volts
    c) 20,000 volts d) 25,000 volts

69. **Generally the gap between the contact breaker points is kept between**
    a) 0 .1 to 0.3 mm b) 0.2 and 0.4 mm
    c) 0.3 and 0.5 mm d) 0 .4 and 0.6 mm

70. **Variation between no load and full load conditions is found to approx. at**

a) 80 rpm b) 91rpm.
c) 100 rpm d) 110 rpm

71. **The economical diesel fuels have got octane rating varying from**

a) 30 to 50 b) 30 to 60
c) 20 to 30 d) 25 to 30

72. **The temperature produced on the power stroke of an engine can be as high as**

a) 1200°C b) 1400°C
c) 1600°C d) 1800°C

73. **Average yearly load on the tractor should not exceed its rated load up to**

a) 70% b) 72%
c) 7 5 % d) 78%

74. **The depreciation method which reduces the value of a machine by equal amount each year during the useful life is known as**

a) Estimated method
b) Constant percentage method
c) Compound interest method
d) Straight line method

75. **The use ful life of a machine is also affect edbyther ate of which machine becomes**

a) Defunct b) Non operational
c) Replaceable d) Obsolete

76. **The glow plugs must be inspected and decarbonized during**

a) Daily maintenance b) Weekly maintenance
c) Periodic maintenance d) Seasonal maintenance

77. **Tractor in small field gives**

a) High efficiency b) Low efficiency
c) Normal efficiency d) Optimum efficiency

78. **A battery must be turned over to a charging station in winter if discharged by**

a) 50% b) 40%
c) 30% d) 25%

79. **Transmission of oil should be changed after every working hours about**

a) 550 hours b) 650 hours
c) 70 hours d) 850 hours

80. **Engine oil should be changed and replaceable type filters should be changed after even working hours about**

a) 420 hours b) 320 hours
c) 220 hours d) 120hours

81. **Front and rear wheel axle bearings may be washed and replaced once in**

a) Four years b) Three years
c) Two years d) One year

82. **If the spark plugs have become found it should be cleaned after working hours of**

a) 25 hours b) 30 hours
c) 40 hours d) 50 hours

83. **Under dusty conditions, air cleaner oil is changed after every working hours of**

a) 6 hours b) 8 hours
c) 10 hours d) 12 hours.

84. **Hydraulic system is mainly used for operating**

a) One point hitch b) Two point hitch
c) Three point hitch d) Four pond hitch.

85. **Hydraulic pump delivers a pressure of about**

a) $125kg/cm^2$ b) $130kg/cm^2$
c) $140kg/cm^2$ d) $150kg/cm^2$

86. **The tie rod connects the front wheel spindle arms and is actua ted by the steering shaft through**

a) Chain link b) Gear link
c) Tension link d) Drag link

87. **Under normal operating conditions, the bearings are filled with the proper type of lubricant after every hours of working about**

a) 6 hours b) 8 hours
c) 10 hours d) 12hours

88. **Depending upon the size of the tractor the width o fthe tracks varies for**

a) 10-50 cm b) 15.50 cm
c) 20 - 25cm d) 25-50 cm

89. **Inflation pressure in the rear wheels of the tractor varies between**

a) 0.5 to 1.2 kg/cm^2 b) 8 to 1.2 kg/cm^2
c) 2 to 1.5 kg/cm^2 d) 1.5 to 20 kg/cm^2

90. **As a result of spinning sometimes one of the drive wheels of a tractor happens to meet a**

a) Hard ground b) Soft ground
c) Muddy gourd d) Wet ground.

91. **For hard facing base alloys used are**

a) High carbon steel b) High carbon iron
c) Carbon steel d) Stainless steel

92. **Hardening is a process of heating a metal to**

a) 400°C b) 500°C
c) 600°C d) 700°C

93. **To obtain softness and ductility, metals art**

a) Tempered b) Hardened
c) Enamelled d) Hard faced

94. **With high frequency electric current hardening is achieved and is called as**
 a) Case hardening b) Induction hardening
 c) Surface hardening d) Metal hardening

95. **Nails are available in various lengths ranging from**
 a) 1.00 to 15 cm b) 1.25 to 15 cm
 c) 1.5 to 15 cm d) 2.5 to 15 cm

96. **A bearing designed to resist axial load is known as**
 a) Ball bearing b) Bush bearing
 c) Thrust bearing d) Journal bearing

97. **The bearings which resist radialloads isknown as**
 a) Ball bearing b) Bush bearing
 c) Thrust bearing d) Journal bearing

98. **Aluminum is a very light metal which resists**
 a) Scratch b) Moisture
 c) Temperature d) Corrosion

99. **Soft center steel is used on**
 a) Plow share b) Plow bottom
 c) Cultivated shovel d) Sweap

100. **Flats belts are available in various widths ranging from**
 a) 1.0 to 50 cm b) 1.5 to 50 cm
 c) 2 to 50 cm d) 2.5 to 50 cm

101. **The datum adopted for India is**
 a) MSL at Mumbai b) MSL at Madras
 c) MSL at Karachi d) None

102. **The B.M. established by the survey of India is known as**
 a) Permanent B.M b) G.T.S.B.M.
 c) Arbitrary BM d) None

103. **The surface of still water is considered to be**

a) Level b) Horizontal
c) Smooth d) None

104. **The surface tangential to level surfaces said to be a**

a) Vertical surface b) Horizontal surface
c) Gourd surface d) None

105. **The line of collimation and axis of telescopic should**

a) Coincide b) be parallel
c) be perpendicular d) None

106. **Which there is relative movement between cross hairs and staff reading, is it known as**

a) Parallax b) Co lunation error
c) Refraction d) None

107. **The staff reading taken a point of known elevation is termed as**

a) FS reading b) BS reading
c) IS reading d) None

108. **Internal focusing telescope is focused moving the**

a) Convex lens b) Double concave leas
c) Plano covers lens d) None

109. **By arithmetical check we can ensure the accuracy of**

a) Field work
b) Calculation
c) Both field work and calculation
d) None

110. **The operation of leveling to determine the elevation between two points is known as**

a) Simple leveling b) Fly leveling
c) Differential leveling d) None

111. B. M. fixed at the end of day's work is called as

a) Permanent B.M. b) Temporary B.M.

c) Arbitrary B.M. d) None

112. The operation of leveling from finishing point to the starting point at the end of days work is knownas

a) Check leveling

b) Reciprocal leveling

c) Cross sectional of leveling

d) None

113. The operation of leveling across any river is termed as

a) Profile leveling b) Reciprocal leveling

c) Compound leveling d) None

114. The operation of leveling from any B.M. to the starting point of at of any project is known as

a) Longitudinal leveling g

b) Fly leveling

c) Continues leveling

d) None

115. To eliminate collimation error the leveling instrument must be placed.

a) Near to BS

b) Near FS

c) Exactly midway between BS and FS

d) None

116. The real image of the object is formed

a) In the plane of cross hairs

b) At the center of eye piece

c) At the centre of telescope

d) None

117. **The diaphragm is filled**
   a) At the centre of telescope
   b) In front of eye piece
   c) In front of object grass
   d) None

118. **The tangent to longitudinal surface of the bubble tube is known as the**
   a) Axis of bubble
   b) Centre line of bubble
   c) Profile of bubble
   d) None

119. **The sensitiveness of the bubble is directly related to**
   a) Length of bubble tube
   b) The radius of curvature
   c) The cross section of bubble tube
   d) None

120. **The most important requirement for leveling is that the**
   a) Horizontal cross hair should lie in a plane perpendicular to vertical axis.
   b) Axis of the bubble tube should be perpendicular to the vertical axis.
   c) Line of sight should be parallel to the axis of the bubble tube.
   d) None

121. **Leveling of dumpy level is done to make**
   a) The vertical axis truly vertical
   b) The line sight truly horizontal
   c) The optical axis truly horizontal
   d) None

122. **The height of instrument (H.I) in leveling is**
   a) Height of telescopic axis above ground
   b) Elevation of line of sight with respect to a datum.
   c) Elevation of line of sight with respect MSL
   d) None of the above

123. **The surface perpendiculars to the direction of the gravity is**

a) Horizontal surface b) Level surface

c) Horizon d) None of the above

124. **A bubble tube with division 2 mm and a radius of 10 m has the sensitivity of about**

a) 40" b) 80"

c) 20" d) 2"

125. **The sensitivity of a bubble tube can be increased by**

a) Increasing the diameter tube

b) Increasing the radius of curvature of tube

c) Increasing the length of tube

d) Increasing length of divisions

126. **With a rise in temperature, the sensitivity of the bubble tube**

a) Decreases b) Increases

c) Remains unaffected d) All Above

127. **On a turning point**

a) Only back sight is taken

b) Only fore sight is taken

c) Both back sight and fore sight are taken

d) An intermediate sight is taken

128. **The rise and fall method provides an arithmetic check on**

a) Back sights and fore sights

b) Intermediate sights

c) Both (a) & (b)

d) Neither (a) and (b)

129. **The following type of leveling can not be done with dumpy level**

a) Differential leveling b) Reciprocal leveling

c) Trigonometric leveling d) Profile leveling

130. **The internal focusing lens is a**

a) Convex lens b) Concave lens
c) Plano convex d) Plano concave

131. **The Philadelphia rod is type of**

a) Solid staff b) Folding staff
c) Telescopic staff d) None of the above

132. **While leveling in an undulating terrain it is preferable to set up the instrument on**

a) The top of submit b) The bottom of valley
c) One side of slope d) All Above

133. **Which leveling of steep slope it is preferable to setup the instrument successively along a**

a) Straight line b) Zig Zag line
c) curved line d) All Above

134. **Select the correct statement**

a) The first reading after the setting up the instrument is a foresight
b) If the foresight is grater than back sight it represents a rise
c) The height instrument method provides a check on intermediate sights.
d) The staff reading is increased when it is fitted.

135. **Which of the following statements is incorrect?**

a) The contours lines are closed curves
b) In steep slopes spacing of contour is small
c) Contour interval on map can vary
d) The direction of steepest slope is at right angles to contour

136. **An imaginary line lying throughout the ground surface and having constant inclination to horizontal is a**

a) Contour line b) Ridge line
c) Contour gradient d) None of the above

137. **The direct method of contouring is**

a) A quick method
b) An accurate method
c) Used for very large area
d) None of the above

138. **The contour interval is**

a) Directly proportional to scale of map
b) Inversely proportional to scale of map
c) Large for flat grounds
d) Small for hill areas

139. **Select the correct statement**

a) Closed contours with higher inwards represent a depression
b) The contour lines crossing a valley have higher values on the convex sides
c) Profile leveling is done for determining contours
d) The horizontal and vertical scales in longitudinal section are generally equal

140. **In reciprocal leveling the error which is not fully eliminated due to**

a) Earths curvature b) Inclined line of sight
c) Refraction d) Both a & c

141. **Constant vertical distance between two adjacent contours is called as**

a) Contour gradient b) Contour interval
c) Horizontal equivalent d) None of the above

142. **If the level tube is out of adjustment by an angle e, when the telescope is rotated through 180° it becomes.**

a) 4e b) 3e
c) 2e d) – e

143. **The line joining points of equal elevation is known as.**
   a) Horizontal line b) Contour line
   c) Level line d) All Above

144. **The vertical distance between two adjacent contour lines is called as a.**
   a) Contour gradient b) Vertical equivalent
   c) Vertical interval d) All Above

145. **The line formed along intersection of the ground surface and a level surface is known as a**
   a) Contour line b) Watershed line
   c) Level line d) All Above

146. **A contour line intersects a ridge line or valley line.**
   a) Obliquely b) Perpendicularly
   c) Vertically d) All Above

147. **Contour interval for particular map is**
   a) Kept constant b) Made variable
   c) Made irregular d) None

148. **When contour line touches at particular zone it indicates a**
   a) Level surface b) Vertical cliff
   c) Horizontal surface d) None

149. **When lower values are inside a loop it indicates a**
   a) High ground b) Level ground
   c) Depression d) All Above

150. **When higher values are inside loops it indicates**
   a) Hill b) Pond
   c) Slopping ground d) All Above

151. **The contour interval is inversely proportional to**
   a) Steepness b) Extent of the area
   c) Scale of map d) All Above

152. **When contour interval is fixed 0.25 to 0.50 m it indicate**

a) A steep slope b) Flattish slope
c) Almost level ground d) None

153. **The alignments of Highways are generally taken along the.**

a) Ridge line b) The valley line
c) Across the contour line d) All Above

154. **When contours different elevation cross each other it indicates an**

a) Vertical cliff b) Saddle
c) Overhanging cliff d) All Above

155. **The horizontal distance between two consecutive contours is termed as**

a) Horizontal interval b) Horizontal equivalent
c) Horizontal contours d) None

156. **When consecutive contour lines runs close together it indicates a**

a) Steep slope b) Flatter slope
c) Vertical surface d) All Above

157. **As per land utilization capability classification (LUCC) the suitable land for cultivation is**

a) Class I to IV b) Class I to V
c) Class V to VII d) None of the above

158. **Most economical cross section of grassed waterway is**

a) Trapezoidal b) Triangular
c) Parabolic d) All of the above

159. **Grassed waterways are designed to carry the peak runoff rate for**

a) 10 years recurrence interval
b) 15 years recurrence interval

c) 20 years recurrence interval
d) None of the above

160. **Grassed waterways are designed by average velocity of flow as**

a) 1.0 m/s b) 1.5 to 2.0 m/sec
c) 2.5m / sec d) None of the above

161. **Velocity of flow of grassed waterway is calculated by equation**

a) $^{1}/n$ RT! S ½ b) $^{1}/n$ R ½ S $^{2}/3$
c) $^{1}/n$ R $^{2}/3$ S ¼ d) None of the above

162. **Diversion ditches are constructed**

a) Across the slope b) Along the slope
c) At the field boundary d) All of the above

163. **Function of diversion ditch is to**

a) Intercept the runoff b) Divert runoff
c) Both a & b d) None of the above

164. **Broad based terrace are constructed in the form of**

a) Steps or benches b) Channels
c) Ridges and channels d) None of theabove

165. **Techniques of Broad base terrace is developed by**

a) U.K. b) India
c) U.S.A d) U.S.S.R.

166. **Alignment of broad base is carried out more or less**

a) Along the contour b) Across the contour
c) Across the slope length d) All of the above

167. **Broad base terrace is also known as**

a) Bench terrace b) Channel terrace
c) Drainage type terrace d) None of the above

168. **Drainage typeterrace are not suitable for**
   a) Humid zones b) Semi arid zones
   c) Tropical zones d) All of the above

169. **Absorption type terraces are most suitable to the**
   a) Humid regions b) Semi arid zones
   c) Tropical zones d) All of the above

170. **Bench terracing is also called**
   a) Hill farming
   b) Staircase farming
   c) Erosion control farming
   d) None of the above

171. **The width of bench terrace with level top is equal to the inverse of**
   a) Degree of slope land b) Depth of cut
   c) Runoff rate d) None of the above

172. **For same land slope the width of level bench terrace is always**
   a) Greater than bench width with forward lateral slope
   b) Less than bench width with forward lateral slope
   c) Equal to the bench width with forward lateral slope
   d) None of the above

173. **The VI (Vertical Interval) of bench terrace with level top should be equal to**
   a) 1/2 of the depth of cut
   b) 1/4 of the depth of cut
   c) 2 times depth of cut
   d) None of the above

174. **Catch area of V shape gully is**
   a) More b) Less
   c) Zero d) None of the above

175. **Drop structures are used to control**

a) V shape gullies b) U shaped gullies
c) Both a & b d) None of the above

176. **Drop structures are designed to handle peak (runoff) discharge rate for return period of**

a) 25 to 50 years b) 10 to 15 years
c) 50 years d) 100 years

177. **Chute spillways are used at**

a) Mild slope b) Steepslope
c) Both a & b d) None of the above

178. **Drop inlet spillways are constructed for**

a) Water retention along with gully bed
b) Controlling water velocity
c) Creating silt deposition
d) Both a & b

179. Rational method is suitable to compute peak runoff rate from

a) Large watersheds b) Small watersheds
c) Medium watersheds d) None of the above

180. **Rational formula states that Q = $^{1}/360$ CIA, where**

a) I in mm/h b) A in ha
c) L in cm/h d) a & b

181. Runoff coefficient is ratio of

a) Annual rainfall to annual runoff
b) Runoff to rainfall
c) Runoff to rainfall intensity
d) All of the above

182. **USLE is**

a) An empirical equation
b) A theoretical solution

c) Both a & b
d) None of the above

183. **High silt contents make a soil to**
a) Less erodible b) More erodible
c) More cohesive d) None of the above

184. **Rain fall intensity is ratio of**
a) Time to rainfall amount
b) Rainfall amount to time
c) Rainfall amount to falling velocity
d) Both b & c

185. **As flow velocity increases from 1m/s to 2 m/s the transportation of soil particles increases as**
a) 32 times b) 16 times
c) 10 times d) 4 times

186. **Bench terraces are constructed at slope gradient of**
a) 5 to 10 % b) 10 to 20 %
c) Less than 5 % d) All of the above

187. **Surface sealing problem is less on**
a) Steep slope b) flat land
c) Vegetated surface d) Both a & c

188. **Clay soil particles more difficult to detach by runoff but easier to**
a) Transport b) Deposit
c) Disintegrate d) Both b & c

189. **For land slope 0 to 9 % the horizontal spacing between terrace should be**
a) Not more than 210 m b) More than 210 m
c) 210 m d) None of the above

190. **Planimeter measures**

a) Area b) Density
c) Land roughness d) All of the above

191. **Area below a unit by hydrographis equal to**

a) 1 cm rainfall excess b) 10 cm direct runoff
c) a & b d) None of the above

192. **A DRH (Direct Runoff Hydrograph) is the plot of**

a) Direct runoff rate and time
b) Rainfall excess and time
c) Rainfall depth and time
d) Both b & c

193. **A unit hydrograph is derived**

a) A unit depth rainfall excess
b) A unit depth runoff
c) A unit depth of rainfall
d) All of the above

194. **Basin lag is the distance measured from the**

a) Out let to opposite side of basin centroid along main stream
b) Outlet to remote point of drainage basin
c) Both a & b
d) None of the above

195. **"Curve number" method predict the**

a) Peak runoff rate
b) Direct runoffs / rainfall excess
c) Sub surface runoff
d) All of the above

196. **Time required for translated the flow from most remotest point of drainage basin to the outlet is called as**

a) Overland flow time b) Time of concentration
c) Rainfall duration d) None of the above

197. **'Contour trenching' is performed to**
   a) Control the land slide form the hill face
   b) Conserve the moisture
   c) Both a & b
   d) None of the above

198. **Contour trench is done mainly in**
   a) Low rainfall areas for moisture conservation
   b) High rainfall areas for moisture conservation
   c) Arid zone for wind erosion control
   d) All of the above

199. Level bench terrace are
   a) Irrigate terrace
   b) Hill type bench terrace
   c) Orchard type bench terrace
   d) All of the above

200. **Narrow width bench terrace are**
   a) Orchard bench terrace
   b) Level bench terrace
   c) Both a & b
   d) None of the above

201. **Vertical Interval of bench terrace is equal to**
   a) Twice of bench width b) Width of bench terrace
   c) Both a & b d) None of the above

202. **Bund constructed on contour of land is**
   a) Contour bund b) Graded bund
   c) Both a & b d) None of the above

203. **Graded bunds are useful in areas of**
   a) Medium rainfall b) Medium to high rainfall
   c) Very less rainfall d) All of the above

204. **Freeboard is provided to accommodates the embankment height against**
   a) Over topping
   b) Sliding
   c) Overturning
   d) All of the above

205. **Location of emergency spillway in embankment pond should be at**
   a) Maximum flood level during expected frequency
   b) Minimum flood level during expected frequency
   c) Both a and b
   d) None of the above

206. **Watershed is**
   a) Drainage basin contributing runoff to single outlet
   b) Unit of area extent in which hydrologic cycle completes
   c) Hydrological entity
   d) All of the above

207. **Seepage line is also called us**
   a) Phreatic line
   b) Saturation line
   c) Both a & b
   d) None of the above

208. **Control of gully erosion can be achieved by use of**
   a) Spurs
   b) Wooden plank dams
   c) Retaining wall
   d) None of the above

209. **Side slope of bund is function of**
   a) Bunds height
   b) Angle of repose of fill material
   c) Both a & b
   d) None of the above

210. **Conduction heat transfer is quantified by**
   a) Fourier's law
   b) Laplace law
   c) Burke – Plummer equation
   d) Blake- Kozney equation

211. **Specific heat, coefficient of viscosity and thermal conductivity are related in**

a) Thermal diffusivity b) Prandtl number
c) Schmidt number d) Froude number

212. **Heat transfer coefficient in natural convection increases with increasing**

a) Boit Number b) Nusselt number
c) Grashof number d) None of the above

213. **Constant rate of drying is directly proportional to**

a) Convective heat transfer coefficient
b) Latent heat of vaporization
c) Wet bulb temperature
d) None of the above

214. **Freeze drying time is directly proportional to the ------ of the material Being dried.**

a) Thickness b) Square of the thickness
c) Cube of thickness d) Fourth power of thickness

215. Decimal reduction time in microbial destruction is inversely proportional to

a) Z value b) Universal gas constant
c) Initial concentration d) Reaction rate

216. **Losses of nutrients during heating of food products are guided by ----- order reaction.**

a) Zero c) First
c) second d) none of the above

217. **Dry ice is known as**

a) Solidified water kept in moisture free environment
b) Solidified Freon-12
c) Solidified carbon dioxide
d) Solidified nitrogen

218. **Most spoilage and pathogenic bacteria those contaminate food materials have water activity in the range of**

a) 0.91 to 0.99 b) 0.86 to 0.90
c) 0.81 to 0.85 d) 0.75 to 0.80

219. **Under the most ideal conditions a bacterium may reproduce itself as often as**

a) Every 20 to 30 seconds b) Every 20 to 30 minutes
c) Every 20 to 30 hours d) Every 2 to 3 days

220. **The orange color of tomato is due to**

a) Chlorophyll A b) Anthocyanins
c) Xanthans d) Lycopene

221. **Presumptive test is found positive for**

a) E. Coli
b) B. Coli
c) Lactobacillus bulgaricus
d) Clostridium botulinum

222. **The differential speed of rolls of wheat mill is**

a) 2.0:1 b) 2.5:1
c) 3.0:1 d) 3.5:1

223. **One tonne of refrigeration is equivalent to the heat required to melt one tone ofice in**

a) 6 h b) 12 h
c) 18 h d) 24 h

224. **Calorific value of rice husk is approximately**

a) 1.25×10^4 KJ/kg b) 2.33 x 104 KJ/kg
c) 4.65×10^4 KJ/kg d) 4.94×10^4 KJ/kg

225. **If a fat globule of 4 mm in diameter moves upwards at a velocity of 1.06 mm/h the velocity of fat globule which is twice this size will be**

a) 2.12 mm/h b) 3.18 mm/h
c) 4.24 mm/h d) 5.30 mm/h

226. **Chhana balls are expanded due to**
   a) Protein denaturation
   b) Differential vapour pressure
   c) Shortage of lactose
   d) Osmotic pressure

227. **Energy required to break a drop of liquid into small droplets will depend Mainlyon**
   a) The surface tension of the liquid
   b) The viscosity of the liquid
   c) The density of the liquid
   d) The heat capacity of the liquid

Answers

1	c	2	c	3	d	4	c	5	b	6	c	7	c	8	c	9	b	10	b
11	d	12	c	13	d	14	d	15	b	16	b	17	d	18	d	19	a	20	a
21	a	22	b	23	a	24	c	25	c	26	b	27	b	28	b	29	d	30	c
31	d	32	d	33	c	34	d	35	b	36	b	37	b	38	d	39	c	40	c
41	c	42	c	43	c	44	b	45	c	46	c	47	b	48	c	49	d	50	c
51	c	52	a	53	d	54	d	55	b	56	c	57	c	58	d	59	b	60	a
61	d	62	c	63	b	64	c	65	c	66	a	67	c	68	c	69	c	70	c
71	b	72	c	73	c	74	d	75	d	76	d	77	b	78	d	79	c	80	d
81	d	82	d	83	c	84	c	85	d	86	d	87	c	88	c	89	b	90	b
91	b	92	c	93	b	94	b	95	b	96	c	97	d	98	d	99	b	100	d
101	c	102	b	103	a	104	b	105	a	106	b	107	a	108	b	109	b	110	b
111	c	112	c	113	a	114	b	115	b	116	c	117	a	118	b	119	a	120	b
121	c	122	a	123	b	124	b	125	a	126	b	127	a	128	c	129	c	130	c
131	b	132	c	133	c	134	b	135	d	136	d	137	c	138	b	139	b	140	b
141	c	142	b	143	c	144	b	145	c	146	a	147	b	148	a	149	b	150	c
151	a	152	a	153	b	154	a	155	c	156	b	157	a	158	c	159	a	160	b
161	a	162	d	163	b	164	c	165	a	166	b	167	b	168	c	169	a	170	a
171	b	172	b	173	a	174	b	175	c	176	c	177	b	178	d	179	c	180	d
181	c	182	a	183	a	184	b	185	a	186	b	187	d	188	a	189	a	190	a
191	a	192	b	193	b	194	b	195	b	196	b	197	c	198	a	199	a	200	a
201	a	202	a	203	b	204	a	205	a	206	a	207	a	208	b	209	b	210	a
211	b	212	c	213	a	214	b	215	d	216	b	217	c	218	a	219	b	220	c
221	a	222	b	223	d	224	a	225	c	226	b	227	c						

❑❑❑

Chapter 4

Agroforestry

1. **Agroforestry technique of growing arable crops between hedge rows of tree and shrubs is known as**

   a) Alley cropping b) Misxed cropping
   c) Relly cropping d) Interculture cropping

2. **Area of nursery should be .... per cent of total plantation area.**

   a) 0-3.0 b) 1-5
   c) .5-2.5 d) 5-8

3. **The Chipko movement is started by**

   a) Engler b) Sunderlal Bahuguna,
   c) Medha Padkar d) None of these

4. **Extensive planting of trees to increase forest cover is called**

   a) Agroforestry b) Afforestation
   c) Deforestation d) Social forestry

6. **Bee-keeping is known as**

   a) Apiculture b) A broad leaved tree species is
   c) Mongo tree d) Banyan tree

7. **Deforestation refers to**

   a) Planting of trees
   b) Cultivation of crops
   c) Disappearance of forests
   d) Depletion of forests

8. **Zoological Survey of India is at**
   a) Delhi b) Mumbai
   c) Kolkata d) Chennai

9. **The book silent spring was written by**
   a) Madhav Gadgil b) Anil Agarwal
   c) Rachel Carson d) E.O.Wilson

10. **An introduced plant species that are neither native nor indigenous are known as Supplementary plant**
   a) Complementary plant b) Exotic plants
   c) Toxic plant d) Native Plants

11. **Agriculture crops forest trees are the component of**
   a) Agrisilviculture b) Agri pasture
   c) Agrisilvi pasture d) Agriculture

12. **World wildlife week is observed during**
   a) First week of September
   b) Last week of October
   c) Third week of October
   d) First week of October

13. **MoEF means**
   a) Ministry of Forest and Energy
   b) Ministry of Environment and Forests
   c) Ministry of Fuel and Energy,
   d) Management of Environment and Forestry

14. **King of forest is**
   a) Mango b) Sagoon
   c) Teak d) Banyan

15. **Land area from water drains to given point is known as**
   a) Command area b) Gross area
   c) Watershed d) Irrigated area

16. **Forests grow in high rain fall areas are**
    a) Ever green forests b) Temperate forests
    c) Conifers d) Tundra

17. **………………… trees shed their leaves during winter and hot summer.**
    a) Ever green b) Coniferous
    c) Deciduous d) Shola

18. **In india, research work on agroforestry was intiated in 1960 at**
    a) Jhansi b) Matura
    c) Varanasi d) Lackhnow

19. **Forestry + agronomy + fishery + animal husbandry are the components of**
    a) Silvi-aqua-pastroral b) Silvi-pastroral-agro
    c) Agro-api-silvi –aqua d) Agro- silvi-aqua-pastroral

20. **Home garden agroforestry system is practiced in**
    a) Dry land b) Humid land
    c) Sub humid low land d) Irrigated land

21. **In India, tropical evergreen forests occur in**
    a) Himachal Pradesh b) Madhya Pradesh
    c) Assam d) Tamil Nadu

22. **Which one is nature's cleaner?**
    a) Consumers
    b) Producers
    c) Decomposers and Scavengers
    d) Symbionts

23. **Early growth stage of tree species requires**
    a) Light pruning b) Training
    c) Shed pruning d) Cutting

24. **The injury to leaves or twigs by strong wind refers to**
    a) Wind blast b) Leaf blast
    c) Leaf fall d) Leaf break

25. **Value of NPK in teak is**
    a) 105:84:46 b) 120:90:50
    c) 100:80:60 d) 90:84:50

26. **What is desertification?**
    a) Conversion of forests into desert
    b) Conversion of croplands into desert,
    c) Conversion of grassland into desert
    d) All of the above

27. **What is deep ecology?**
    a) Protecting nature for its own sake
    b) Ecology of deep sea
    c) Ecology of forest
    d) Ecology of caves

28. **World forest day is celebrated on**
    a) 22 April b) 21 March
    c) 22 March d) 15 October

29. **The pruning of dead branches means**
    a) Green pruning b) Dead pruning
    c) Dry pruning d) Dead leaf pruning

30. **A set of organisms that resemble one another in appearance and behavior is called a ____________**
    a) Exons b) Prions
    c) Species d) Guilds

31. **Temperate forests occur in India in**
    a) Indo-Gangetic plains b) Himalayas
    c) Eastern India d) Southern peninsula

32. **The tree species spread parallel to ground in order to enhance more shoots is known as**
    a) Bending b) Training
    c) Pruning d) Cutting

33. **The pruning of living branches is**
    a) Green pruning b) Leaf pruning
    c) Green leaf pruning d) Both A &B

34. **The function of agroforestry are**
    a) Productive b) Protective
    c) Development of forest d) Both A &B

35. **Plant species grow in stagnated water are known as**
    a) Aquatic plants b) Rice
    c) Water Plants d) Immersed plants

36. **suitable agroforestry system for waste land is**
    a) Silvi –pastoral b) Agro- silvi
    c) Agro-silvi- pastoral d) Both A&C

37. **What is edge species?**
    a) The species which occur abundantly in the ecotone
    b) Species on the verge of extinction
    c) Rare species
    d) Endemic species

38. **What is desertification?**
    a) Conversion of forests into desert
    b) Conversion of croplands into desert
    c) Conversion of grassland into desert
    d) All of the above

39. **Sweet potato should not grown with**
    a) Banana b) Mango
    c) Cluster bean d) Lichi

40. Which of the following was not a feature of blandongdiensten system?

a) Rent exemption

b) Free labour and buffaloes for cutting forests

c) Right to cultivate forest land

d) Free labour for transportation

41. Fee paid by people from one village to another

a) Grazing tax b) Rent

c) Devsari d) Mand

42. Soybean and green should not be grown with

a) Cluster bean b) Pointed gourd

c) Banana d) Sugarcane

43. Taungya word comes from Burma is

a) 1850 b) 1900

c) 1869 d) 1960

44. Plantation grow as linear strip of land on the side of public road canal and railway line are known as

a) Avenue plantation b) Road Plantation

c) Strip Planting d) Canal planting

45. state having highest area of wasteland is

a) Rajasthan b) Madhya Pradesh

c) Uttar Pradesh d) Tamil nadu

46. Which of the following was not a method adopted by Saminists?

a) Lying down on the land

b) Refusing to pay taxes and fines

c) Refusing to perform labour

d) Flogging the Dutch

47. Which of the following is a new development in forestry?

a) Conservation of forests

b) Collecting timber

c) Scientific forestry

d) Keeping communities away from forests

48. **System includes both traditional and modern land use system where tree is managed with crops and animal rearing is known as**

a) Agroforestry b) Crop science

c) Animal Science d) Modern croping

49. **The scorched earth followed by policy was**

a) Japanese in Java b) British in India

c) French in Vietnam d) Dutch in Java

50. **Pricking time of eucalyptus tereticornis day after sowing**

a) 20-30 b) 10-20

c) 30-35 d) 20-50

51. **The method of cultivation usually followed in hills by tribal peoples is**

a) Shifting cultivation b) Really cropping

c) Mixed cultivation d) Inter cultivation

52. **Which among the following is the main reason for survival of some dense forests in India?**

a) Conservationists

b) Environmentalists

c) Scientific forestry

d) Villages protected them as sacred grooves

53. **Sacred grooves are known as**

a) Sarnas, Kan, Rai, Devarakudu

b) Dhya, Penda, Bewar

c) Nevad, Jhum, Podu

d) Siadi, Semur, Chena Tavi,

54. **The pruning of roots for the purpose of encouraging compact fibrous root system is**

a) Root pruning b) Root cutting
c) Root training d) Root nourishment

55. **In which system the trees are grown around the fish pond**

a) Aqua forestry b) Farm Pond
c) Fish pond d) Agro-forestry

56. **Tree having toxic effect on soil and associated crops**

a) Eucalyptus b) Tamarinds
c) Teak d) Both A&B

57. **Tree or shrub maintained for the purpose of shelter from wind sun show drift etc. are knows as**

a) Silviculture b) Horticulture
c) Sericulture d) Agro-forestry

58. **The practice of agri silviculture was began in**

a) Burma b) Usa
c) China d) Maxico

59. **The tree plants known as protein bank are**

a) Acacia nilotica b) Albizialebbeck
c) Sebania grandiflora d) All of the above

60. **Ramsar Convention refers to the conservation of ___________**

a) Deserts b) Wetlands
c) Agriculture lands, d) Forest land

61. **The area to which a species is biologically adapted to live is known as ________**

a) Niche b) Habitat
c) Succession d) Community

62. **Cultivation of agricultural crops with horticultural crops and forest tree species is knows as**

a) Agri- horti- silviculture b) Horti-agri-silviculture

c) Silvi-agri-horticulture d) Silviculture

63. **Nitrogen fixing trees are**

a) Acacia nilotica b) Albizialebbeck

c) Prosopis d) All of the above

64. **Rearing of fish with forest tree is known as**

a) Aqua silviculture b) Aqua forestry

c) Aqua pastoral d) Agro aqua

65. **Mango should not grown with**

a) Sugarcane b) Wheat

c) Jowar d) All of the above

66. **pricking time of albizialebbeck days after sowing is,**

a) 55 b) 35

c) 10 d) 60

67. **A tree shed its leaves yearly at the end of growing season and go for some rest is known as**

a) Deciduous tree b) Evergreen tree

c) Tropical trees d) Sub-tropical trees

68. **A group of plants with in a species genetically adapted to a particular habitat are known as**

a) Ecotype plants b) Habitat plants

c) Wetland plants d) Dry palnts

69. **Growing of woody plant with agricultural crop and live stock on the same unit of land is known as**

a) Agroforestry b) Animal husbandry

c) Combined farming d) Mix farming

70. **Pomegranate+ dhaman grass is the example of**

a) Horti–pastoral b) Agro-silvi

c) Horti–silvi d) Both B&C

71. **Plant growing in desert are known as**

a) Xerophytes b) Dry plants

c) Eco-phytic d) Desert type

Answer

1 a	2 C	3 b	4 b	5 a	6 b	7 c	8 c	9 c	10 c
11 a	12 d	13 b	14 c	15 c	16a	17c	18a	19 d	20 c
21 c	22 c	23 a	24 b	25 a	26d	27a	28b	29 c	30 c
31 b	32 a	33 a	34 d	35 c	36a	37a	38d	39 c	40 c
41 c	42 a	43 a	44 a	45 a	46d	47a	48a	49 d	50 b
51 a	52 d	53 a	54 a	55 b	56d	57a	58a	59 d	60 b
61 b	62 a	63 d	64 a	65 d	66c	67c	68b	69 a	70 a
71 c									

❑❑❑

Chapter 5

Agronomy

1. **First agriculture minister of independent India.**
   a) T. C. Patel b) K. C. Pant
   c) Rafi Ahmed Kidwai d) None of these

2. **Who is consider as president of ICAR**
   a) President of india
   b) Central Agriculture Minister
   c) Director General of ICAR
   d) Governor of Delhi

3. **Net sown area of India**
   a) 306 mha b) 142 mha
   c) 328 mha d) 95 mha

4. **Round revolution related to**
   a) Potato b) Tomato
   c) Onion d) Fruit production

5. **ICRISAT is situated at**
   a) Hyderabad (1957) b) New Delhi (1972)
   c) Manila (1960) d) Hyderabad (1972)

6. **Food and Agriculture Organization (FAO) is located at**
   a) Geneva b) Hyderabad
   c) Rome d) New York

7. **Food and Agriculture Organization (FAO) is establishment in**
   a) 1945 b) 1995
   c) 1905 d) 1929

8. **Norman E. Borlaug got Nobel Prize in the field of**
   a) Peace b) Physics
   c) Chemistry d) Economics

9. **APEDA stands for**
   a) Agricultural and Processed Food Products Exports Development Agency
   b) Agricultural and Processed Food Products Exports Development Authority
   c) Agricultural Products Exports Development Authority
   d) Agricultural Products Exports Development Agency

10. **ISOPOM (Integrated Scheme of Oilseed, Pulses, Oil Palms and Maize) wasstarted in**
   a) 1965 b) 2000
   c) 2008 d) 2004

11. **Pineapple disease is associated with which crop?**
   a) Pineapple b) Sugarcane
   c) Groundnut d) Sugarbeet

12. **Which green manure crop contain high nitrogen**
   a) Dhaincha b) Sunhemp
   c) Cowpea d) Guar

13. **Chicory and Cuscuta are associated with**
   a) Berseem and lucern
   b) Lucern and berseem
   c) Sorghum and pearmillet
   d) None of these

14. **The cropping intensity of maize-potato-wheat-blackgram**
   a) 500% b) 350%
   c) 400% d) 300%

15. **The term Leaf Area Index (LAI) was given by**
   a) Watson b) Donald
   c) Buckman d) Wiley

16. **The term Harvest Index (HI) was given by**
    a) Donald (1963) b) Watson (1955)
    c) Arnon (1939) d) Donald (1968)

17. **CAZRI was establishment in**
    a) Hyderabad (1959) b) Hyderabad (1957)
    c) Cuttack (1946) d) Junagrah (1961)

18. **Mixed farming is referred as**
    a) Growing of crop together by mixing the seed
    b) Growing of cash crops and food grains together
    c) Crop production combined with dairy farming at farm
    d) Growing of fruits and vegetable mixed

19. **For getting 15 % advantage from intercropping system, the LER should be**
    a) 0.85 b) 1.15
    c) 2.30 d) 0.70

20. **National Institute of Organic Farming is located at**
    a) Ghaziabad b) Hyderabad
    c) New Delhi d) Modipurram

21. **Which is primary source of agriculture**
    a) Fertilizer b) Seed
    c) Pesticide d) Land

22. **Optimum time of sowing is example of**
    a) Low cost input b) Non monetary input
    c) High cost input d) None of these

23. **Crop grown to attract the insect**
    a) Trap crop b) Catch crop
    c) Relay crop d) Break crop

24. **In India which state declared as fully organic state**
    a) Karnataka b) Madhya Pradesh
    c) Sikkim d) Arunachal Pradesh

25. **APEDA and IFOAM are associate with**
    a) Organic farming b) Natural farming
    c) Cropping system d) Intensive farming

26. **Example of indeterminate crop is**
    a) Sorghum b) Rice
    c) Cotton d) All of these

27. **Drip method of irrigation can be recommended to**
    a) Rice b) Sorghum
    c) Groundnut d) Cotton

28. **Who has given term green revolution**
    a) William Gaud b) M. S. Swaminathan
    c) N. E. Borlaug d) B. P. Pal

29. **Which year is declared as Sustainable Tourism for development**
    a) 2007 b) 2011
    c) 2015 d) 2017

30. **Pseudocereal is**
    a) Duram wheat b) Buckwheat
    c) Macroni wheat d) Bread wheat

31. **Which part of the leaf is present in wheat but not in the grassy weeds such as *Phalaris minor* and *Avena fatua*?**
    a) Auricle b) Ligule
    c) Both auricle and ligule d) None of these above

32. **In India how many crops got minimum support price.**
    a) 19 b) 20
    c) 24 d) 27

33. **The organization related to fixing minimum support price of Agriculturalproduct is**
    a) NITI b) CACP
    c) Ministry of agriculture d) APEDA

34. **Minimum support price Fixed by**
    a) State government b) Central government
    c) Ministry of agriculture d) Government of India

35. **Minimum support price should be declared**
    a) Before harvesting of crop
    b) Before sowing of crop
    c) Before flowering of crop
    d) None of these

36. **Which organization related to refinancing of agriculture is**
    a) APMC b) NABARD
    c) CACP d) APEDA

37. **In which fertilizer not getting subsidy.**
    a) Ammonium sulphate b) Urea
    c) MAP d) SSP

38. **National Agriculture policy was started in which year.**
    a) 2000 b) 2004
    c) 2014 d) 2016

39. **Which practices are related to sugarcane?**
    a) Wrapping and tying b) Propping
    c) Trashing d) All of the above

40. **Per hectare consumption of $N+P_2O_5+K_2O$ in India is about**
    a) 90 kg/ha b) 100 kg/ha
    c) 140 kg/ha d) 120 kg/ha

41. **When LER is more than one, intercropping is**
    a) Benefit
    b) Loss
    c) Nighters benefit nor loss
    d) None of these

42. **Central Soil and Water Conservation Research and Training Institute was established in**

a) Pusa (1905) b) Deharadun (1974)

c) Hyderabad (1972) d) Mumbai (1982)

43. **The duration of photoperiod of short day plant is**

a) <10 hrs b) 12-14 hrs

c) >14 hrs d) None of these

44. **The massive hole in Ozone layer over the Antarctica was first discovered in**

a) 1960 b) 1985

c) 1990 d) 1995

45. **Which gas is 20 times more global warming potential than CO_2**

a) SO_2 b) CH_4

c) NH_3 d) N_2O

46. **Base temperature for wheat is**

a) 4.5 ^0C b) 21 ^0C

c) 10 ^0C d) 35 ^0C

47. **To reach the earth's surface the rays of the sun take about**

a) 18 minutes b) 12 minutes

c) 8 minutes d) 30 minutes

48. **What is the result when industrial chemical combine with condensing vapour?**

a) Global warming b) Green house effect

c) Base rain d) Acid rain

49. **World Environment day celebrated on**

a) 23th March b) 5th June

c) 22th April d) 16th September

50. **The term green house effect was given by**
    a) Fourlier b) Linnaeus
    c) Arrhenius d) Tyndall

51. **Zero tillage system was first used during 1950 for pasture renovation in**
    a) USA b) India
    c) Japan d) France

52. **Water requirement for land preparation with pudling is around**
    a) 50 mm b) 200 mm
    c) 300 mm d) 500 mm

53. **Zero tillage is mostly practiced under.................. cropping sequence**
    a) Wheat-wheat b) Wheat-rice
    c) Rice-rice d) Rice-wheat

54. **FIRB technology is planting of**
    a) Wheat b) Pearlmillet
    c) Potato d) Groundnut

55. **BBF technology is planting of**
    a) Wheat b) Pearlmillet
    c) Potato d) Groundnut

56. **Set line planting is practiced in Gujarat for cultivation of**
    a) Groundnut b) Cotton
    c) Both a and b d) Sugarcane

57. **Ultra narrow cultivation is practiced in**
    a) Cotton b) Maize
    c) Groundnut d) Rice

58. **CRIDA defined depth of deep tillage is**
    a) 5-10 cm b) 15-25 cm
    c) 25-30 cm d) 50-70 cm

59. **International rice research institute was established during**

a) 1980 b) 1946
c) 1947 d) 1960

60. **Central Rice Research Institute (CRRI) is located in**

a) Cuttack b) Hyderabad
c) New Delhi d) Bangalore

61. **In Dapog method of nuesery quantity of seeds required for one squaremeter area is**

a) 1 kg b) 3 kg
c) 5 kg d) 4 kg

62. **Days required for rice seedling to transplanting under SRI methods**

a) 10-12 b) 12-14
c) 14-16 d) 25-30

63. **Tillering in rice begins at the stage of**

a) 1-2 leaf b) 2-3 leaf
c) 3-4 leaf d) 4-5 leaf

64. **Zinc deficiency of rice was first reported by**

a) Dr. R. S. Paroda b) Dr. M. S. Swaminathan
c) Dr. B. P. Pal d) Dr. Y. L. Nene

65. **Akiochi in rice is associated with**

a) Sulphide toxicity b) Potassium deficiency
c) Iron toxicity d) None of these

66. **The nursery are required to provide seedling for transplanting one hectarearea in roughly**

a) $1/10^{th}$ b) $1/4^{th}$
c) $2/5^{th}$ d) None of these

67. **Golden rice is a rich source of**

a) Vitamin B b) Vitamin A
c) Vitamin C d) Vitamin D

68. **The N uptake of rice increased from 50 Kg N/ha under control (no nitrogen) to 70 kg N/ha with application of 40 ka N/ha. What will be the apparent N recovery?**

a) 40 % b) 50 %
c) 60 % d) 80 %

69. **Seed rate of hybrid rice**

a) 5 kg/ha b) 15 kg/ha
c) 25 kg/ha d) 45 kg/ha

70. **Green revolution was started in which five year plan**

a) First five year plan b) Second five year plan
c) Third five year plan d) Fourth five year plan

71. **Optimum seed rate require for normal sown wheat is**

a) 90-100 kg/ha b) 100-125 kg/ha
c) 125-150 kg/ha d) 150-250 kg/ha

72. **For late sown wheat, seed rate is**

a) 90-100 kg/ha b) 100-125 kg/ha
c) 125-150 kg/ha d) None of these

73. **Crown root initiation stage in dwarf wheat is at**

a) 15-20 DAS b) 20-25 DAS
c) 25-30 DAS d) 30-35 DAS

74. **In Karnal bunt disease black powder gives a foul smell due to presence of**

a) Trimethylamine b) Methylamine
c) Both a and b d) None of these

75. **Among the following cereals the highest average productivity in India is**

a) Rice b) Maize
c) Wheat d) Barley

76. **If a farmer has provision of only one irrigation in wheat, at what stage he should irrigated the crop?**
    a) Tillering b) Jointing
    c) Flowering d) CRI

77. **Which of the following crop consume maximum fertilizer in India**
    a) Wheat b) Rice
    c) Maize d) Potato

78. **Which of the following cereals is largely used in malting and brewing industries?**
    a) Wheat b) Oats
    c) Barley d) Pearlmillet

79. **Which of the followings crop is most hardy against salinity and sodocity**
    a) Barley b) Wheat
    c) Maize d) Sorghum

80. **In USA which crop is known as yellow gold**
    a) Soybean b) Maize
    c) Berseem d) Sarson

81. **Most critical stage for irrigation in maize**
    a) Tasseling b) Silking
    c) Panicle initiation d) Both a and b

82. **Maize crop planted at a spacing of 50 x 20 cm will have plants populationper hectare is**
    a) 50,000 b) 1,00,000
    c) 1,25,000 d) 1,50,000

83. **Which of the followings is the most important insect of sorghum?**
    a) Grass hopper b) Shoot fly
    c) Termites d) White grub

84. **The toxic limit of HCN in sorghum is**
   a) 200 ppm b) 300 ppm
   c) 400 ppm d) 500 ppm

85. **Soil sickness is caused by**
   a) Maize b) Wheat
   c) Pearlmillet d) Sorghum

86. **Which of the following states is having maximum area and production of Pearlmillet**
   a) Maharashtra b) Rajasthan
   c) Gujarat d) Madhya Pradesh

87. **Ergot is an important disease of**
   a) Fingermillet b) Kodomillet
   c) Pearlmillet d) Foxtailmillet

88. **The success of transplanting is maximum in**
   a) Pearlmillet b) Maize
   c) Sorghum d) None of these

89. **Which among the following has maximum protein content**
   a) Sorghum b) Maize
   c) Barley d) Pearlmillet

90. **Presence of which acid on leaves of gram is beneficial for stomach ailments**
   a) Malic acid b) Oxalic acid
   c) Both a and b d) None of these

91. **Indian Institute of Pulses Research (IIPR) in located at**
   a) Bharatpur b) Kanpur
   c) Udaipur d) Modipuram

92. **All India Coordinated Research Project on chickpea was**
   a) 1967 b) 1981
   c) 1993 d) 1999

93. **Most commonly and widely cultivated pulse in India is**
    a) Chickpea b) Redgram
    c) Pea d) Lentil

94. **Which of the following operation is spatially followed in chickpea**
    a) Desuckering b) Propping
    c) Retting d) Nipping

95. **Seed rate of Desi and Kabuli gram is**
    a) 25-50 kg/ha and 50-75 kg/ha
    b) 50-75 kg/ha and 75-100 kg/ha
    c) 75-100 kg/ha and 125-150 kg/ha
    d) 125-150 kg/ha and 150-200 kg/ha

96. **For Desi and Kabuli gram row spacing is kept at**
    a) 30 cm and 45 cm b) 45 cm and 75 cm
    c) 75 cm and 90 cm d) 30 cm and 90 cm

97. **First hybrid of pigeon pea released in 1991 was**
    a) ICPH-8 b) COH-1
    c) COH-2 d) PPH-4

98. **The varieties of ICPL series were developed from**
    a) TNAU, Coimbatore b) IARI, New Delhi
    c) ICRISAT, Hyderabad d) BHU, Varanasi

99. **Which among the following is termed as golden gram**
    a) Chickpea b) Greengram
    c) Blackgram d) Frenchbean

100. **The first mutant variety of green gram was released in**
    a) 1980 b) 1984
    c) 1989 d) 1994

101. **Which Rhizobium species is suitable for pea and lentil**

a) Rhizobium leguminoserum

b) Rhizobium Phaseoli

c) Rhizobium meliloti

d) Rhizobium japonicum

102. **Which of the following known as chickling pea.**

a) Chickpea b) Lathyrus

c) Lentil d) Frenchbean

103. **BOAA (â- oxayl amino alanine) a free amino acid is found in which of the following crops.**

a) Frenchbean b) Fieldbean

c) Chickpea d) Lathyrus

104. **The safe limit of BOAA in lathyrus is**

a) <0.2 % b) <0.5 %

c) <0.7 % d) <0.9 %

105. **Which among the following pulse is drought tolerant**

a) Lentil b) Frenchbean

c) Lathyrus d) Pea

106. **Which among the following pulse have neurotoxin**

a) Lathyrus b) Lentil

c) Frenchbean d) Kulti

107. **The excess use of which among the following pulse causes lathyrism**

a) Lentil b) Lathyrus

c) Frenchbean d) None of these

108. **The low neurotoxin variety of lathyrus is**

a) Bio L-212 b) Pusa 24

c) LSD 3 d) LSD 6

109. **Nitrogen requirnment of rajmash is**

a) 60-80 kg/ha b) 80-100 kg/ha
c) 100-120 kg/ha d) 120-140 kg/ha

110. **Which among the followings is most sensitive to frost and sodocity**

a) Lentil b) Gram
c) Fieldpea d) Rajmash

111. **Which among the following pulse are more tolerant to heavy rainfall**

a) Blackgram b) Greengram
c) Cowpea d) Gram

112. **Which among the followings is known as black-eyed pea**

a) Blackgram b) Cowpea
c) Greengram d) Horsegram

113. **Which among the following pulse crop used as a pulse, fodder and green manure crop**

a) Moong b) Urd
c) Cowpea d) Pea

114. **Which among the following pulse is highly drought resistant**

a) Cowpea b) Blackgram
c) Greengram d) Mothbean

115. **Which of the following pulse is valuable for patients suffering from urinary trouble**

a) Horsegram b) Chickpea
c) Lathyrus d) Lentil

116. **Main industrial use for guar is**

a) Gum b) Textile
c) Explosives d) Cosmetics

117. **Vegetable oil is obtained in India from how many cultivated oilseedcrops**

a) 3 b) 5
c) 7 d) 9

118. **Total number of oilseed crop cultivated in India**

a) 3 b) 5
c) 7 d) 9

119. **Among the following nine vegetable oilseed crops, which is non-edible oil crops**

a) Linseed b) Castor
c) Both a & b d) None of these

120. **All India Coordinated Research Project on Oilseed was started in**

a) 1960 b) 1962
c) 1967 d) 1970

121. **King of vegetable oilseeds in India is considered to**

a) Soybean b) Groundnut
c) Mustard d) Sesame

122. **In groundnut, seed dormancy is a problem in**

a) Bunch type b) Spreading type
c) Both a & b d) None of these

123. **The calorific value of groundnut is considered is**

a) 349 per 100 g b) 375 per 100 g
c) 400 per 100 g d) 625 per 100 g

124. **The spacing of spreading type of groundnut is**

a) 30 cm x 10 cm b) 45 cm x 10 cm
c) 40 cm x 10 cm d) 22 cm x 10 cm

125. **The plant population for bunch and runner type of groundnut isgenerally kept as**

a) 3.33 lakh/ha and 2.22 lakh/ha, respectively

b) 2.22 lakh/ha and 3.33 lakh/ha, respectively

c) 4 lakh/ha and 2 lakh/ha, respectively

d) None of these

126. Deep ploughing of soil in a groundnut filed should be normally avoided because it

a) Delays harvesting

b) Delays germination

c) Makes harvesting difficult

d) Delays peg and pod development

127. **Dormancy of groundnut can be broken by**

a) Ethrel solution b) Sulphuric acid

c) Auxin d) gibberalic acid

128. **Gypsum in groundnut is applied**

a) At the time of sowing b) At the time of podding

c) At the time of pegging d) At the time of peak flowering

129. **Intercultural operation in groundnut is avoided at the crop stage**

a) Flowering b) Seedling

c) Pegging d) Pod formation

130. **The permissible limit of aflatoxin content in groundnut os**

a) 6 ppb b) 18 ppb

c) 20 ppb d) 32 ppb

131. **The safe moisture limit for storage of groundnut is**

a) 5-6 % b) 8-9 %

c) 11-12 % d) 12-13 %

132. **The highest limit of free fatty acid in groundnut for its good keeping quality is**

a) 1.0 % b) 2.5 %
c) 4 % d) 5 %

133. **The pops in groundnut is due to deficiency of**

a) Calcium b) Nitrogen
c) Phosphorus d) Potassium

134. **Most preferred source of phosphorus in groundnut is**

a) DAP b) Rock phosphate
c) SSP d) None of these

135. **Maximum consumption of edible oil in India is**

a) Mustard b) Coconut
c) Groundnut d) Sesame

136. **'00' Brassicas is also known as**

a) Canola b) Taramira
c) Mustard d) Toria

137. **In canola the erucic acid in seed is less than .... of total fatty acid**

a) 2 % b) 3 %
c) 4 % d) 5 %

138. **Fruit of mustard is**

a) Capsule b) Pod
c) Siliqua d) Caryposis

139. **The test weight of seeds of Brassica varies between**

a) 3 to 5 g b) 6 to 8 g
c) 9 to 11 g d) 12-14 g

140. **Parasitic weed associated with mustard is**

a) Striga b) Cuscuta
c) Loranthus d) Orobanche

141. **Cheapest adulteration in mustard is**
   a) Maxican poppy seeds b) Opium poppy seeds
   c) Linseed seeds d) Niger seeds

142. **Sangam is a variety of which oil seed crop**
   a) Mustard b) Taramira
   c) Toria d) Yellow sarson

143. **The crop of group of rapeseed and mustard (except taramir a) belong to the family**
   a) Cruciferae b) Leguminosae
   c) Compositae d) Euphorbiaceae

144. **The Technology Mission on Oilseed was launched in**
   a) 1976 b) 1986
   c) 1996 d) 2006

145. **The Technology Mission on Oilseed, Pulses and Maize was started inwhich five year plan**
   a) Seven five yaer plan b) Ninth five year plan
   c) Tenth five year plan d) Twelfth five year plan

146. **Which among the following is a major oilseed crop in the world**
   a) Groundnut b) Rapeseed and mustard
   c) Soybean d) Sunflower

147. **The oil and protein content in soybean is**
   a) 20 and 40%, respectively
   b) 40 and 20%, respectively
   c) 30 and 60%, respectively
   d) 60 and 30%, respectively

148. **Which rhizobium species is suitable for soybean**
   a) Rhizobium meliloti
   b) Rhizobium phaseoli
   c) Rhizobium leguminasorum
   d) Rhizobium japonicum

149. **Which food is designated as 'boneless' meat**

a) Soybean b) Cowpea
c) Pigeonpea d) Mothbean

150. **Inflorescence of sunflower is called**

a) Arrow b) Capitulum
c) Silk d) Head

151. **Which one is a dwarf and short duration variety of sunflower**

a) Morden b) DSH-1
c) Jwalamukhi d) Surya

152. **Number of bee hives required per hectare to pollinate sunflower is**

a) 2 b) 3
c) 4 d) 5

153. **Sunflower act as an indicator plant to diagnose the deficiency of**

a) Boron b) Iron
c) Nitrogen d) Phosphorus

154. **Rancidity in sunflower oil is caused by the process of**

a) Reduction b) Oxidation
c) Esterification d) Nitrification

155. **Sesame belong to family**

a) Compositae b) Linaceae
c) Pedaliaceae d) Solanaceae

156. **Phyllody is a serious disorder of**

a) Linseed b) Castor
c) Toria d) Sesame

157. **Which of the following oilseed crop play an important role inreducing blood cholesterol**

a) Groundnut b) Sesame
c) Sunflower d) Safflower

158. **Carthamus tinctorius is a botanical name of**
   a) Linseed b) Sesame
   c) Sunflower d) Safflower

159. **Poly Unsaturated Fatty Acid (PUF A) content highest in oil of**
   a) Groundnut b) Sesame
   c) Sunflower d) Safflower

160. **Linseed intemperate countries is mainly grown for**
   a) Oil b) Sticks
   c) Fibre (Flax) d) Seed

161. **Rayungans is seed material associated with**
   a) Rice b) Elephant grass
   c) Pearl millet d) Sugarcane

162. **The planting of sugarcane by trench method**
   a) Improves tillering b) Reducing lodging
   c) Improve germination d) Increase water use efficiency

163. **Which portion of sugarcane stalk is preferred for sowing**
   a) Top one-third to half portion
   b) Top one-fourth portion
   c) Bottom one-fourth portion
   d) Bottom one third to half portion

164. **Which of the following is called as noble cane**
   a) Saccharum sinense b) Saccharum officinarum
   c) Saccharum robustum d) Saccharum barberi

165. **Which sugarcane variety is known as first wonder cane in India**
   a) CO 740 b) CO 1148
   c) CO 1253 d) CO 419

166. **Which is second wonder cane variety**
   a) CO 740 b) CO 1148
   c) CO 1253 d) CO 419

167. **Which growth stage of sugarcane is most critical for irrigation**
   a) Formative stage b) Flowering stage
   c) Maturity stage d) None of these

168. **Which of the following is a chemical clarificant of sugarcane juice**
   a) Lime water b) Sodium carbonate
   c) Alum d) All of these

169. **Generally sugarcane flowers in**
   a) 9-12 months b) 6-9 months
   c) 12-15 months d) None of these

170. **First intergenetic sugarcane hybrid was developed in**
   a) 1996 b) 1936
   c) 1916 d) 1946

171. **Highest productivity of sugarcane is in**
   a) UP b) Maharashtra
   c) Tamil Nadu d) Karnataka

172. **Which of the following is a major disease of sugarcane**
   a) Red rot b) Wilt
   c) Mosaic d) Ratoon stunting disease

173. **Which is the most common pre emergence herbicide used insugarcane**
   a) Pendimethalin b) Fluchloralin
   c) Atrazine d) Paraquat

174. **Sugarbeet belong to family**
   a) Linaceae b) Chenopodiaceae
   c) Solanaceae d) Liliaceae

175. **Sugarbeet is an indicator plant for**
   a) Mo b) Na
   c) Mn d) Mg

176. **The ratio of cotton seed to lint is**
   a) 1:3 b) 1:2
   c) 2:1 d) 3:1

177. **Seed cotton means**
   a) Seed of cotton b) Seed+lint
   c) Lint d) None of these

178. **In defuzzing process for 10 kg cotton, the amount of sulphuric acidrequired is**
   a) 1 lit b) 2 lit
   c) 3 lit d) 4 lit

179. **"Boll shedding" in cotton is due to**
   a) Plant over loaded with bolls
   b) Inhibitory effect of ABA
   c) Heavy dressing of urea
   d) Frequent irrigation

180. **Delinting of cotton seeds help in**
   a) Grading the seed
   b) Killing of hibernating boll worm larva and disease pathogens
   c) Quicker germination
   d) All of these

181. **The world first interspecific hybrid of cotton is**
   a) Hybrid-4 b) Varalaxmi
   c) Dhanlaxmi d) Savita

182. **Seed rate of hybrid cotton is**
   a) 2 kg/ha b) 4 kg/ha
   c) 6 kg/ha d) 8 kg/ha

183. **Weight of one bale of cotton**
   a) 160 kg/ha b) 170 kg/ha
   c) 180 kg/ha d) 200 kg/ha

184. **Which of the following is a physiological disorder in cotton**
   a) Tirak
   b) Red leaf disease
   c) Shedding of fruiting parts
   d) All of these

185. **The process of dipping of jute bundles in water is called**
   a) Steeping b) Extraction
   c) Retting d) None of these

186. **Ribboning is associated with which crop**
   a) Cotton b) Jute
   c) Linseed d) Sunhemp

187. **Retting is a process by which**
   a) The poor fibre stalk sorted out
   b) The fibre in the bark get loosened and separated from woody stalk
   c) The good quality fibre is sorted out
   d) Fibre are differentiated into various retted groups

188. **The fibre purpose, jute may be harvested at any time**
   a) After pod formation b) After flowering
   c) Before pod formation d) Before flowering

189. **Area required for nursery of tobacco for sowing in one hectare area**
   a) $1/10^{th}$ of hectare b) $1/20^{th}$ of hectare
   c) $1/100^{th}$ of hectare d) $1/1000^{th}$ of hectare

190. **In tobacco, correct chronological order of practice**
   a) Topping-desuckering-priming
   b) Desuckering-topping-priming
   c) Priming-desuckering-topping
   d) Priming-topping desuckering

191. **Which potassic fertilizer should not be applied in tobacco**
   a) Potassium chloride
   b) Potassium sulphate
   c) Potassium sulphate nitrate
   d) None of these

192. **Combustibility of tobacco leaves is governed by**
   a) Chlorine b) Potassium
   c) Iodine d) Calcium

193. **Nicotine is produced in tobacco**
   a) Stem b) Roots
   c) Leaves d) None of these

194. **Concentration of nicotine in tobacco is maximum in**
   a) Stem b) Roots
   c) Leaves d) None of these

195. **Seed rate of tobacco for transplanting**
   a) 0.5-1.0 kg/ha b) 1-1.5 kg/ha
   c) 2-3 kg/ha d) 5-6 kg/ha

196. **Thiurea is used to break dormancy in**
   a) Groundnut b) Rice
   c) Sunflower d) Potato

197. **Aerial stem of potato is known as**
   a) Haulms b) Stolon
   c) Both a & b d) None of these

198. **Central Potato Research Institute is located at**
   a) Pantnagar b) New Delhi
   c) Bangalore d) Shimla

199. **Which herbicide is used as a desiccant in potato**
   a) Glyphosate b) Diuron
   c) Simazine d) Paraquat

200. **Seed rate of true potato seed (TPS) is around**
   a) 50 g/ha b) 150 g/ha
   c) 250 g/ha d) 350g/ha

201. **The largest cropping system in India**
   a) Rice-wheat b) Rice-pulses
   c) Rice-rice d) Wheat-rice

202. **One of the major limitation in rice-wheat cropping system is**
   a) Delayed wheat sowing b) Delayed rice sowing
   c) Early rice sowing d) Early wheat sowing

203. **Cropping intensity of India is around**
   a) 130 % b) 135 %
   c) 140 % d) 145 %

204. **Cropping intensity means**
   a) Ratio of net cropped area to gross cropped area x 100
   b) Ratio of gross cropped area to net cropped area x 100
   c) Ratio of number of crops in rotation to period of one rotation x 100
   d) None of these

205. **Khaira disease of rice is diagnosed by**
   a) Necrotic lesions
   b) Dark reddish brown pigmentation
   c) Yellowish spot on leaves
   d) Black pin head point on leaf

206. **Chlorosis is observed in upland rice due to deficiency of**
   a) Sulphur b) Iron
   c) Zinc d) Manganese

207. **Nitrogen is taken by the plant in the form of**
   a) Elemental nitrogen b) Nitrate
   c) Nitrite d) None of these

208. **When emerging leaves of maize show yellow to white bleached bands in lower parts, it shows the deficiency of**
   a) Iron b) Boron
   c) Copper d) Zinc

209. **Which of the following is a nitrification inhibitor**
   a) A.M. (2, amino-4, 6 methyl pyridine)
   b) Thio-urea
   c) Sulphur coated urea
   d) CBC (β-chloro ethyl, trimethyl ammonium chloride)

210. **The bacteria considered most important to bring about the conversion of NH_4^+ to NO_2**
   a) Nitrobacter b) Nitrosomonas
   c) Both a & b d) Azotobacter

211. **Anaerobic environment in paddy soil is responsible for gaseous loss of fertilizer nitrogen by**
   a) Ammonification b) Nitrification
   c) Denitrification d) Volatilization

212. **Symbiotic association of fungal hyphae with plant roots is known as**
   a) Mutual infection b) Lichen
   c) Co-parasitism d) Mycorrhiza

213. **Which of the following strains is used for biological nitrogen fixation in soybean**
   a) Rhizobium leguminoserum
   b) Rhizobium japonicum
   c) Rhizobium meliloti
   d) Rhizobium trifoli

214. **Green ear disease of bajra is suppressed by increasing the concentration of**
   a) Nitrate b) Calcium
   c) Phosphate d) Potassium

215. **Which of the following is non metallic nutrient**

a) Calcium b) Chlorine
c) Boron d) Magnesium

216. **Organic matter can be calculated by multiplying organic carbon by**

a) 1.72 b) 1.82
c) 1.92 d) 1.27

217. **Which of the following nutrient is recommended for oil seed crops along with NPK fertilizers**

a) Boron b) Sulphur
c) Magnesium d) None of these

218. **Which of the following plant nutrients is useful in increasing resistance to disease and insect pest**

a) Calcium b) Phosphorus
c) Nitrogen d) Potassium

219. **Hard fruit of citrus is due to deficiency of**

a) Boron b) Calcium
c) Manganese d) Zinc

220. **Organic carbon (%) in organic matter is**

a) 16 b) 46
c) 50 d) 58

221. **Improper development of wheat inflorescence is due to deficiency of**

a) Iron b) Zinc
c) Boron d) Phosphorus

222. **White bud of maize caused due to deficiency of**

a) Zinc b) Manganese
c) Copper d) Molybdenum

223. **The most deficient micro nutrient in Indian soil**

a) Fe b) Mn
c) Zn d) Cu

224. **Black heart disease in potato is due to deficiency of**
   a) Carbon dioxide b) Oxygen
   c) Nitrogen d) Phosphorus

225. **Which nutrient is a constituent of nitrate reductase and nitrogenaseenzyme**
   a) Iron b) Manganese
   c) Zinc d) Molybdenum

226. **Which field crop consider as poor man's saffron**
   a) Niger b) Safflower
   c) Sunflower d) Sesame

227. **Whenever one irrigation is available, the wheat crop should beirrigated at**
   a) CRI stage b) Tillering stage
   c) Flowering stage d) Milking stage

228. **Water use efficiency is maximum in**
   a) C_3 plants b) C_4 plants
   c) CAM plants d) None of these

229. **One hectare meter is equal to........m^3 of water**
   a) 100 b) 1000
   c) 100000 d) 10000

230. **1 ha cm of irrigation water is equal to**
   a) 1000 lit b) 10000 lit
   c) 100000 lit d) 1000000 lit

231. **World water day is celebrate on**
   a) January, 22 b) March, 22
   c) February, 22 d) April, 22

232. **Most widely used method for soil moisture determination in India**
   a) Gravimetric method b) Gypsum blocks
   c) Tensiometer d) Neutron probe

233. **The unit used to expressed water used efficiency is**

a) Kg/ha b) Kg/ha mm

c) Kg/ha m d) Kg/ha cm

234. **Water requirement of crop is equal to**

a) Consumptive use of water+ application losses

b) Consumptive use of water

c) Consumptive use of water+ application losses+ water needed for special purpose operation

d) All of these

235. **Which instrument is used to measure rate of infiltration**

a) Anemometer b) Infiltrometer

c) Hydrometer d) Hygrometer

236. **Which river basin has highest ground water potential**

a) Godavari b) Narmada

c) Brahmputra d) Ganga

237. **Direct method of soil water is**

a) Tensiometer b) Gravimetric

c) Neutron scattering d) None of these

238. **Corrugation is a type of irrigation**

a) Surface b) Furrow

c) Sub surface d) None of these

239. **What is the pH of pure water**

a) 6.0 b) 6.5

c) 7.0 d) 7.5

240. **Chemical that found to be most suitable as water evaporation inhibitoris**

a) Methyl alcohol b) Ethyl alcohol

c) Acetyl alcohol d) All of these

241. **Water holding capacity maximum in**

a) Sandy soil b) Clay soil
c) Silty soil d) Loamy soil

242. **The ratio between yield and ET is called**

a) Water use efficiency
b) Crop water use efficiency
c) Field water use efficiency
d) None of these

243. **2, 4- D is a ……… herbicide**

a) Selective and systemic b) Non selective
c) Contact d) None of these

244. **Trap crop of striga is**

a) Sunflower b) Cotton
c) Cowpea d) All of these

245. **Which of the following is non selective herbicide**

a) Glyphosate b) Paraquate
c) Diquate d) All of these

246. **Glyphosate is used as**

a) Growth inhibitor b) Seedling and stock inhibitor
c) Pigment inhibitor d) Metabolic inhibitor

247. **Which of the following chemical used for suicidal germination ofstriga seeds**

a) Ethylene b) Bromine
c) Methylene d) Sulphuric acid

248. **Which among the following weeds is known as potassium lovers?**

a) Amaranthus b) Digitaria spp.
c) Chenopodium spp d) None of these

249. **Which among the following weeds has highest N content**

a) Argemone Mexicana b) Eclipta alba
c) Melilotus alba d) Amaranthus viridis

250. **Which among the following is the cheapest method of elimination ofweeds**

a) Flaming b) Soil solarization
c) Non-living mulchs d) Herbicides

Answers

1	c	2	b	3	b	4	a	5	d	6	c	7	a	8	a	9	d	10	d
11	b	12	a	13	a	14	c	15	a	16	d	17	a	18	c	19	b	20	a
21	d	22	b	23	a	24	c	25	a	26	c	27	d	28	a	29	d	30	b
31	a	32	c	33	b	34	d	35	b	36	b	37	a	38	a	39	d	40	c
41	a	42	b	43	a	44	c	45	b	46	a	47	c	48	d	49	b	50	a
51	a	52	b	53	d	54	a	55	d	56	c	57	a	58	c	59	d	60	a
61	b	62	a	63	d	64	d	65	a	66	a	67	b	68	a	69	b	70	c
71	a	72	c	73	b	74	a	75	c	76	d	77	a	78	c	79	a	80	b
81	d	82	b	83	b	84	a	85	d	86	b	87	c	88	a	89	d	90	c
91	b	92	a	93	a	94	d	95	b	96	d	97	a	98	c	99	b	100	a
101	a	102	b	103	d	104	a	105	c	106	a	107	b	108	a	109	b	110	d
111	c	112	b	113	c	114	d	115	a	116	a	117	c	118	d	119	c	120	c
121	b	122	b	123	a	124	b	125	a	126	c	127	a	128	d	129	c	130	a
131	a	132	b	133	a	134	c	135	c	136	a	137	a	138	c	139	a	140	d
141	a	142	c	143	a	144	b	145	d	146	c	147	a	148	d	149	a	150	b
151	a	152	d	153	a	154	b	155	c	156	d	157	d	158	d	159	d	160	c
161	d	162	d	163	a	164	b	165	d	166	a	167	a	168	d	169	a	170	c
171	c	172	a	173	c	174	b	175	b	176	c	177	b	178	a	179	b	180	d
181	b	182	a	183	b	184	a	185	a	186	b	187	b	188	c	189	c	190	a
191	a	192	a	193	b	194	c	195	c	196	d	197	a	198	d	199	d	200	b
201	a	202	a	203	c	204	b	205	b	206	b	207	b	208	d	209	a	210	b
211	c	212	d	213	b	214	d	215	c	216	a	217	b	218	d	219	a	220	d
221	c	222	a	223	c	224	b	225	d	226	b	227	a	228	c	229	d	230	c
231	b	232	a	233	b	234	c	235	b	236	d	237	b	238	b	239	c	240	c
241	a	242	b	243	a	244	d	245	d	246	d	247	a	248	c	249	d	250	a

❑❑❑

Chapter 6

Animal Husbandry

1. **Karan Swiss breed of cattle was developed at**

   a) NDRI Karnal b) NIANP Bangalore
   c) NFDB Hyderabad d) CARI Izatnagar

2. **White scours usually occur in**

   a) Calves b) Buffalo
   c) Cattle d) Sheep

3. **Milk fever in cow occurs due to**

   a) Ca deficiency b) P deficiency
   c) Boron deficiency d) Mg deficiency

4. **Bellowing is usually experienced during**

   a) Early heat b) Low heat
   c) Moderate heat d) High heat

5. **Mastitis is an inflammation of**

   a) Plague b) Udder
   c) Rinderpest d) Bronchitis

6. **Udder Brucellosis is caused by**

   a) Bacteria b) Viral
   c) Fungal d) Protozoa

7. **Milk let down is initiated by the hormone of**

   a) Oxytocin b) Cobalamin
   c) Glycogen d) Albumin

8. **Body temperature of cow is**
   a) 100 F b) 101 F
   c) 102.5 F d) 99.5 F

9. **A cow should be given pregnancy ration after**
   a) 5months b) 6 month
   c) 7month d) 12 month

10. **'Swaichal' is the characteristic of the cattle breed**
   a) Kankrej b) Tharparkar
   c) Ongole d) Mewati

11. **In India more than 70 percent of milk producers are**
   a) Medium farmer b) Small farmer
   c) Marginal farmers d) Both B&C

12. **A female calf born as a twin to a male in cattle is known as**
   a) Dolly b) Pratham
   c) Free martin d) Rennin

13. **Normal dry period of cow is**
   a) 50-53 days b) 70-85 days
   c) 110-120 days d) 60-65 days

14. **A pregnant cow doesn't come in heat due to presence by**
   a) Corpus luteum b) Thiamine and biotin
   c) Colostrum d) Theileria

15. **About 60-80 % of energy needs of ruminants is provided by**
   a) Vitamins b) Minerals
   c) Volatile fatty acids d) Linoleic acid

16. **Animal age can be judged by**
   a) No. of teeth b) Horn rings
   c) Age d) None of these

17. **Anthrax is a disease that is**
    a) Contagious b) Coccidiosis
    c) Aflatoxicosis d) Typhoid

18. **'AMUL' the milk marketing system, follows the principle of**
    a) Co-operative b) Group
    c) Partnership d) None of these

19. **The exotic breed of cattle that produces maximum quantity of milk is**
    a) Frieswal b) Brawn swiss
    c) Holstein Friesian d) Jersey

20. **Amrit mahal cattle is native of**
    a) Mysore b) Guntur
    c) Karachi d) Kutchh

21. **In an adult buffalo, approximate weight of rumen content is**
    a) 20%-40% of its body weight
    b) 30%-45% of its body weight
    c) 20-25% of its body weight
    d) 10-24% of its body weight

22. **Cattle plague another name is..........**
    a) Mastitis b) Rinderpest
    c) Cow pox d) Anthrax

23. **When the cattle have all the incisor of cow's calf visible and ready to use**
    a) 3 month after birth b) 2 month after birth
    c) 6 month after birth d) 1 month after birth

24. **According to one estimate about 1200 working hours are provided by**
    a) Bullocks b) Cattle
    c) Sheep d) Goat

25. **When the cattle have all the incisors present and age is above 5 years it is known as**
    a) Full mouth   b) Plague
    c) Cholera   d) Ranikhet

26. **Cellulose is digested in grazing cattle by**
    a) Gastric bacteria   b) Estrogen hormone
    c) Carbohydrates   d) Calving

27. **Cattle disease transferred to man is………**
    a) Ranikhet   b) Ringworm
    c) Anthrax   d) Typhoid

28. **Cattle and buffalo belong to family**
    a) Bovidae   b) Felidae
    c) Canidae   d) Drove

29. **Castration in male calf is done to produce**
    a) Sheep   b) Cattle
    c) Poultry   d) Bullock

30. **Buffalo milk is a rich source of minerals**
    a) Ca   b) P
    c) Mg   d) Both A&B

31. **Buffalo share in total milk production in India is………**
    a) 70%   b) 65%
    c) 20%   d) 55%

32. **Cattle meat is……….**
    a) Buffin   b) Chevon
    c) Beef and calf-veal   d) Mutton

33. **Breed of cattle mainly used for physical work is known as**
    a) Dairy   b) Draught
    c) Dual purpose   d) None of these

34. **Brushing the hair coat of animals is known as**

a) Grooming b) Rumen
c) Weaning d) All of the above

35. **Longest lactation period of exotic cow breed is ....**

a) Sahiwal b) Holesteinfriesian
c) Jersey d) Frieswal

36. **Longest lactation period of mehsana breed of buffalo is**

a) 325 days b) 318 days
c) 280 days d) 310 days

37. **Brown swiss breed is native of**

a) Holland b) Turkey
c) Spain d) Switzerland

38. **Black quarter is a disease**

a) Buffalo b) Cattle
c) Goat d) Sheep

39. **Bull used for breeding purpose is known as**

a) Shed b) Weight
c) Stud bull d) None of these

40. **Body coat colour of mysore type cattle is........**

a) Grey b) Red
c) Black d) White

41. **Bhadawari buffalo is native of ....**

a) Sindh b) Punjab
c) Agra d) Baroda

42. **Best ph for semen preservation**

a) 6.8 b) 5.5
c) 7.2 d) 8

43. **Best method of milking is.........**

a) Full hand b) Automatic

c) Robotic d) All of the above

44. **Best dual purpose breed of cattle is........**

a) Punjab b) Haryana

c) Sindh d) Mysore

45. **Best crop for silage making is...........**

a) Cowpea b) Sorghum

c) Maize d) Buck wheat

46. **Average heart beats per minute of cow is.........**

a) 55 times b) 70 times

c) 80 times d) 300 times

47. **Average gestation length in cattle is.........**

a) 281 b) 150

c) 310 d) 318

48. **Average fertile life of an egg is.........**

a) 18-20 hr b) 10-12 hr

c) 12-24 hr d) 20-25 hr

49. **Best dairy breed of cow in India is.........**

a) Rathi b) Tharparkar

c) Gir d) Sahiwal

50. **Breed known as spotted mountain cattle is.......**

a) German fleckvieh b) Karan fries

c) Karan swiss d) Jersey

51. **Which of the following is the major silk producing state in India?**

a) Karnataka b) Andhra Pradesh

c) Assam d) Bihar

52. **East coast fever in cattle is caused by**
    a) Theileria b) Bacterial
    c) Protozoa d) Fungal

53. **Disbudding in calf is done at the age of**
    a) 7-10 days b) 15-25 days
    c) 80-90 days d) 40-50 days

54. **Cow pox is**
    a) Buffalo b) Sheep
    c) Cattle d) Goat

55. **Cow and buffalo milk contain an average % of fat**
    a) 5.6 % b) 3.5%
    c) 6.1% d) Both B&C

56. **FMD diseases are caused due to**
    a) Water pollution b) Viral
    c) Fungal d) Bacterial

57. **Anthrax and tapedik diseases are caused due to**
    a) Plague b) bronchitis
    c) Water pollution d) Blue tongue

58. **Conception rate in artificial insemination (AI) is**
    a) 58-75% b) 80-85%
    c) 40-60% d) 60-755

59. **Secretion of udder just after calving is**
    a) Carotene b) Colostrum
    c) Linoleic acid d) Albumin

60. **Detection of heat in animal is easier in type of housing**
    a) Loose housing b) Milking house
    c) Both A&B d) None of the above

61. **Degnala and alkali diseases are caused due to**
   a) Selenium toxicity b) Vitamin E deficiency
   c) Deficiency of biotin d) All of the above

62. **Liquid nitrogen is used in cattle for**
   a) Cryo branding b) Brucellosis
   c) Cross breeding d) Embryo transfer

63. **Crude fiber in roughages is**
   a) $>25\%$ b) $<18\%$
   c) $>18\%$ d) $<12\%$

64. **Corpus luteum starts to regress in**
   a) Secretion of adrenaline b) Diestrus
   c) Rumen d) All of the above

65. **Colostrum is generally fed to**
   a) Young calves b) Buffalo
   c) Lamb d) None of the above

66. **Chemical used for dehorning is**
   a) Sodium benzoate b) Caustic potash
   c) Alcohol d) glyphosate

67. **Colostrum is fed to a new born calf as it provides immunity against diseases due to thepresence of**
   a) Cystic ovaries b) Gamma globulins
   c) Rumen d) Prolactin

68. **Digestible crude protein in legume crop is**
   a) 0.8-1.6 b) 1.5-2.8%
   c) 0.2-0.8% d) 0.5 - 1.0%

69. **Common clinical manifestation of brucellosis in animal is known as**
   a) Abortions b) Colostrum
   c) Docking d) All of the above

70. **Digestible crude protein in legume fodder is**

a) 0.5-1.2% b) 5.2-6.8%
c) 2.5-3.0% d) 1.2-2.2%

71. **Frieswalthe heist milk yielding breed yielded**

a) 20 lit/day b) 60 lit/day
c) 80 lit/day d) 14 lit/day

72. **Rinderpest smallpox are**

a) Viral diseases b) Bacterial diseases
c) Fungal diseases d) All of the above

73. **First milk of parturition cattle is known as**

a) Cobalamin b) Colostrum
c) Lactose d) Carotene

74. **Enzyme used in digestion of milk casein in calves is known as**

a) Selenium b) Elastin
c) Rennin d) All of the above

75. **For silage making fodder should be in rich in**

a) Protein b) Vitamin
c) Carbohydrates d) Minerals

76. **First organized plan to improve rural cattle in India was**

a) Rural development scheme
b) Key village scheme
c) Live hood scheme
d) Both B & C

77. **Expulsion of placenta should be within**

a) 8-10 hr b) 20-30 hr
c) 5-7 hr d) 6-8 hr

78. **Expulsion of fetus in cow is known as**

a) Calving b) Lambing
c) Kidding d) Farrowing

79. **Enzyme to break cellulose is present in**
   a) Valine b) Cow
   c) Ruminants d) Both B& C

80. **During parturition water bag appears in**
   a) Secretion stage b) Ruminants stage
   c) Dilation stage d) All of the above

81. **Native place of kankrej cattle is**
   a) Gujarat b) Alwar
   c) Sindh d) Hissar

82. **Karan fries is cross of**
   a) H.F * Tharparker b) Sahiwal * brown swiss
   c) H.F * sahiwal d) None of these

83. **Intercalving period in normal cows should be**
   a) 18-20 months b) 50-60 months
   c) 10-15 months d) 12-13 months

84. **Gestation period of cattle is**
   a) 150 days b) 148 days
   c) 316 days d) 282 days

85. **Gestation period of buffalo is**
   a) 310 b) 150
   c) 300 d) 280

86. **Grass tetany diseases is caused by**
   a) Mg deficiency b) Boron deficiency
   c) Nitrogen deficiency d) Vitamin E

87. **Hallikar is a breed of**
   a) Buffalo b) Lamb
   c) Cattle d) None of these

88. **Heaviest cattle milch breed of India is**
   a) Tharparkar b) Gir
   c) Sindhi d) Sahiwal

89. **Highest milk producing country in the world**
   a) China b) USA
   c) India d) UK

90. **Highest milk producing state in India is**
   a) Madhya Pradesh b) Andhra Pradesh
   c) Gujrat d) Uttar Pradesh

91. **Highest milk yielder breed of buffalo is**
   a) Bhadawari b) Niliravi
   c) Murrah d) Surti

92. **Hormone for milk secretion is**
   a) Linoleic b) Oxytocin
   c) Prolactin d) None of these

93. **Chromosome number found in cow and goat**
   a) 60 b) 50
   c) 42 d) 48

94. **Holstein Friesian native of**
   a) Holland b) English Channel
   c) Switzerland d) Spain

95. **Holding of milk in cow is due to**
   a) Secretion of oxytocin b) Secretion of adrenalin
   c) Both A&C d) None of these

96. **How much saliva volume secreted in cattle per day**
   a) 200-300 litres b) 150-200 litres
   c) 100-200 litres d) 400-460 litres

97. **The zoonotic disease affecting animals as well as human beings is**
    a) Anthrax b) Marek's
    c) Plague d) All of the above

98. **The disease transmitted through bulls in natural breeding is**
    a) Mastitis b) Cholera
    c) Brucellosis d) Ringworm

99. **The gas used for storage of semen**
    a) Nitrogen b) Hydrogen
    c) Oxygen d) Carbon di oxide

100. **Total teeth of cattle buffalo sheep and goat are**
    a) 32 b) 48
    c) 42 d) 51

101. **The timing gap betweentwo milking is**
    a) 10 hr b) 15 hr
    c) 12 hr d) 20 hr

102. **The world's first in vitro fertilized buffalo calf is**
    a) Surati b) Pratham
    c) Dolly d) Both B& C

103. **The viral disease of cattle and buffalo's ulcers are seen in mouth and interdigital space**
    a) Mastitis b) Anthrax
    c) Foot and mouth d) All of the above

104. **To improve the Indian buffaloes the right policy of breeding is**
    a) Selective breeding b) Cross breeding
    c) Mixed breeding d) All of the above

105. **Toda is a breed of**
    a) Cow b) Buffalo
    c) Sheep d) None of these

106. **The highest milk producing indigenous breed of cattle is**
   a) Gir b) Rathi
   c) Tharparker d) Sahiwal

107. **The incubation period of rinderpest disease in cattle**
   a) 5-8 days b) 2-5 days
   c) 3-7 days d) 10-15 days

108. **The disease is caused by shortage or faulty utilization of sugar in the body is known as**
   a) Marek's b) Ketosis
   c) Cholera d) FMD

109. **The genetic contribution of both the male and female can be utilized simultaneously in the**
   a) Embryo transfer technology
   b) Ovaries transfer technology
   c) Rumen transfer technology
   d) None of these

110. **The breeding worth of dairy cattle is evaluated by**
   a) Selective selection b) Pedigree selection
   c) Cross selection d) All of the above

111. **The colour of fat in milk is dependent on**
   a) Casein b) Carotene
   c) Lactose d) Lecithin

112. **The cattle disease that is non-contagious**
   a) Milk fever b) FMD
   c) Cow pox d) Anthrax

113. **The cattle belong to the genus**
   a) Bos b) Pos
   c) Kos d) Mos

114. **The shape of ovary in buffaloes is**
   a) Ovaries shape b) Round shape
   c) Kidney shaped d) None of these

115. **Tuberculin test is known as**
   a) Allergic test b) Vitamin test
   c) Blood test d) None of these

116. **The breed that have five pinkish white marking and known as panchakalyani**
   a) Zaffrabadi b) Murrah
   c) Nili Ravi d) Surti

117. **The country accommodates highest number of buffalo in the world**
   a) India b) China
   c) USA d) Europe

118. **Angora breed of goat is a native of**
   a) Spain b) Europe
   c) Italy d) Turkey

119. **Act of mating in case of sheep is known as**
   a) Mohair b) Tupping
   c) Pica d) Dehorning

120. **In Indiahow many breeds of goat found**
   a) 30 b) 24
   c) 80 d) 16

121. **Native of jamulapari goat is**
   a) Uttar Pradesh b) Andhra Pradesh
   c) Gujrat d) Hissar

122. **Meat of sheep is**
   a) Chevon b) Mutton
   c) Chicken d) Pork

123. **Merino breed of sheep originated from**

a) Turkey b) Europe
c) Italy d) Spain

124. **Merino breed of sheep is**

a) Fine wool breed b) Dual purpose breed
c) Desert breed d) All of the above

125. **Goat like sheep is**

a) Gaddi b) Nellore
c) Deccani d) Merino

126. **Goat meat is known as**

a) Chevon b) Pork
c) Buffine d) Chicken

127. **Gestation period of goat is**

a) 210 b) 148
c) 150 d) 250

128. **First somatically clone of adult animal in the world is**

a) Dolly b) Pratham
c) Assel d) None of the above

129. **Fine leather prepared from goat skin is**

a) Morocco b) Barbari
c) Jamunapari d) Sirohi

130. **First domestic ruminant animal was**

a) Swine b) Goat
c) Sheep d) Poultry

131. **Duration of estrus in goat is**

a) 36-48 hr b) 40-45 hr
c) 60-65 hr d) 35-40 hr

132. **Dorset horn is native of**
   a) Turkey b) Merino
   c) England d) Spain

133. **Gestation period of sheep**
   a) 180 days b) 160 days
   c) 150 days d) 200 days

134. **Feeding extra concentrate to pregnant sheep is known as**
   a) Steaming up b) Culling
   c) Anorexia d) Alveolus

135. **Cost return ratio of goat should be**
   a) 1:22 b) 1:8
   c) 1:11 d) 1:32

136. **Chokla sheep is native of**
   a) Pakistan b) Karnataka
   c) Hissar d) Rajasthan

137. **Browsing is common feature of**
   a) poultry b) Sheep
   c) Goat d) swine

138. **In India wool production is maximum in**
   a) Andhra Pradesh b) Rajasthan
   c) Uttar Pradesh d) Maharashtra

139. **India's rank first in production of**
   a) Carpet wool b) Milk
   c) Poultry d) Both A&B

140. **Group of goats is known as**
   a) Trip b) Group
   c) Capon d) Wedder

141. **Goat are kept in**
   a) Fold b) Cattle
   c) Rumen d) Colostrum

142. **Heaviest breed of sheep of world is**
   a) Lincoln b) Merino
   c) Rambouillet d) None of the above

143. **Goat tissue vaccine is used against of**
   a) Rinderpest b) Ring worm
   c) FMD d) Typhoid

144. **High quality galicha found from sheep breed is**
   a) Lohi b) Chokla
   c) Deccani d) Bikaneri

145. **Improving the nutritional status of ewes during 3 -4 weeks prior to mating is known as**
   a) Flushing b) Alveolus
   c) Culling d) Casting

146. **Indian goat famous for chevon quality is**
   a) Barbari b) Black Bengal
   c) Jamunapari d) Sirohi

147. **Parturition in ewe is known as**
   a) Farrowing b) Hatching
   c) Lambing d) Kidding

148. **Pashmina is produced from one of breed of goat**
   a) Kashmiri b) Morocco
   c) Alpine d) Angora

149. **Parturition in doe is known as**
   a) Kidding b) Lambing
   c) Farrowing d) Hatching

150. **Estrous cycle of sheep is**
   a) 12-18 days b) 20-30 days
   c) 80-90 days d) 15-18 days

151. **Poor man's cow is known as**
   a) Sheep b) Goat
   c) Cattle d) Buffalo

152. **Removal of tail in sheep is**
   a) Docking b) Culling
   c) Casting d) Dehorning

153. **Native place of sanen**
   a) Turkey b) Europe
   c) Switzerland d) Spain

154. **Stiff lamb disease is caused by deficiency of**
   a) Vitamin E b) Vitamin A
   c) Vitamin C d) None of these

155. **Thc only triple purpose Indian breed of sheep is**
   a) Nellore b) Chokla
   c) Lohia d) Gaddi

156. **The valuable textile fiber commercially known as mohair is produced from goat breed**
   a) Alpine b) Saanen
   c) Anglo nubbin d) Angora

157. **Merino wool's share in world**
   a) 82% b) 80%
   c) 75% d) 58%

158. **Mohair is produced from the breed of goat**
   a) Angora b) Alpine
   c) Saanen d) Anglo nubbin

159. **Most popular fine wool breed of world**
   a) Nellore b) Merino
   c) Rambouillet d) Angora

160. **Nutrient conversion efficiency foe goat is**
   a) 45-71% b) 40-50%
   c) 30-39% d) 50-68%

161. **Estrous cycle of goat is**
   a) 30-35 days b) 20-25 days
   c) 18-21 days d) 10-12 days

162. **The famous Kashmiri shawl is prepared from**
   a) Sirohi b) Barbari
   c) Capra d) All of the above

163. **The lightest body weight goat is**
   a) Saanen b) Anglo Nubian
   c) Angora d) Toggenburg

164. **The breed known as milk queen of goat**
   a) Saanen b) Anglo Nubian
   c) Angora d) Toggenberg

165. **The cross bred sheep is**
   a) Nilgiri b) Lohi
   c) Nellore d) Chokla

166. **The species having maximum number of total solids in milk is**
   a) Goat b) Cattle
   c) Buffalo d) Sheep

167. **The disease transmitted by biting insect is**
   a) Blue tongue b) Leucosis
   c) Rinderpest d) Mastitis

168. **The breed of goat known as jersey of goat**

a) Nubian b) Saanen

c) Angora d) Alpine

Answers

1	a	2	a	3	a	4	a	5	b	6	a	7	a	8	b	9	c	10	a
11	d	12	c	13	d	14	a	15	c	16	a	17	a	18	a	19	c	20	a
21	c	22	b	23	d	24	a	25	a	26	a	27	c	28	a	29	d	30	d
31	d	32	c	33	b	34	a	35	c	36	a	37	d	38	b	39	c	40	a
41	c	42	a	43	a	44	b	45	c	46	a	47	a	48	c	49	d	50	a
51	a	52	a	53	a	54	c	55	d	56	a	57	c	58	a	59	b	60	a
61	a	62	a	63	c	64	b	65	a	66	b	67	b	68	d	69	a	70	c
71	d	72	a	73	b	74	c	75	c	76	b	77	d	78	a	79	d	80	c
81	a	82	a	83	d	84	d	85	a	86	a	87	c	88	b	89	c	90	d
91	c	92	c	93	a	94	a	95	b	96	c	97	a	98	c	99	a	100	a
101	c	102	b	103	c	104	a	105	b	106	d	107	c	108	b	109	a	110	b
111	b	112	a	113	a	114	c	115	a	116	c	117	a	118	d	119	b	120	b
121	a	122	b	123	d	124	a	125	b	126	a	127	b	128	a	129	a	130	b
131	a	132	c	133	c	134	a	135	c	136	d	137	c	138	b	139	d	140	a
141	a	142	a	143	a	144	b	145	a	146	b	147	c	148	a	149	a	150	d
151	b	152	a	153	c	154	a	155	c	156	d	157	b	158	a	159	b	160	a
161	c	162	c	163	d	164	a	165	a	166	d	167	a	168	a				

❑❑❑

Chapter 7

Biotechnology

1. **The sugar residues are covalently joined by**
   a) 5′-> 3′ phosphodiester bonds
   b) H_2 bonds
   c) Phosphorylated ions
   d) Covalent bonds

2. **The bases in nucleic acids govern which one of the following process that underpins the essential biological process?**
   a) Replication b) Recombination
   c) Gene expression d) All the above

3. **Nucleotide is made up of**
   a) Sugar b) Phosphate
   c) Nitrogenous base d) All the above

4. **Which one of the following base is present both in DNA and RNA?**
   a) Adenine b) Guanine
   c) Cytosine d) All the above

5. **The diameter of the double helix is**
   a) 20 Å b) 34 Å
   c) 3.4Å d) Varies from one point to another

6. **The number of bases in A form DNA per turn of the helix is**
   a) 20 b) 10
   c) 34 d) 3.4

7. **Which one of the following phosphate moiety involved in 5′ ->3′ phosphodiester bond formation?**
   a) α b) β
   c) γ d) All the above

8. **DNA or RNA polymerases are catalyze the 5′ -> 3′ phosphodiester bond formation and the byproduct of this reaction is**
   a) Pyrophosphate b) Pyrophosphatase
   c) Pyrosequences d) Pyrates

9. **Chargaff's rules state that**
   a) A=T
   b) G=C
   c) Complementarily base paring
   d) All the above

10. **The most stable nitrogenous bases in double helix DNA is**
   a) G:C b) A:T
   c) A:C d) T:G

11. **Who has shown that replication is semi-conservative event?**
   a) Mathew Meselson and Franklin Stahl
   b) Watson and Crick
   c) Delbruck and Korenberg
   d) None of the above

12. **The enzyme which unwinds DNA ahead of replication fork is**
   a) DNA helicase b) DNA ligase
   c) DNA polymerase d) DNA primase

13. ***Taq*DNA polymerase, widely used in PCR, is homologous to *E. coli* pol I but lacks**
   a) 3″ → 5′ exonuclease activity
   b) 5″ → 3′ exonuclease activity
   c) 5″ → 3′ polymerase activity
   d) 3″ → 5′ polymerase activity

14. **During replication, the functional complex primosome is formed by**

a) Primase b) Helicase

c) Both a and b d) α –Polymerase

15. **Enzymes which digest nucleic acid by hydrolyzing phosphodiester bonds are called as**

a) Nucleases

b) Deoxyribo nucleases (if substrate is DNA)

c) Ribonucleases (if substrate is RNA)

d) All the above

16. **Which of the following is true?**

a) DNA polymerase recognizes 'ori' in parental DNA and unwinds the double helix at these sites so that DNA replication can occur.

b) Helicases recognize replication fork and unwinds the double helix at replication fork so that DNA replication can occur

c) Helicases recognize origin of replication in parental DNA and unwinds the double helix at these sites so that DNA replication can occur

d) None of the above

17. **If the parental DNA strand, 3′ GGCATATTCGCTGCAGT 5′, is used as a template DNA strand the newly synthesized, antiparallel strand would be as follows**

a) 3′CCGTATAAGCGACGTCA 5′

b) 5′ CCGTATAAGCGACGTCA 3′

c) 5′ TGACGTCGCTTATACGG 3′

d) 3′ GGCATATTCGCTGCAGT 5′

18. **The sequence of one strand of DNA is 5′-AGTCGACGA-3′. What would be the 5′ to 3′ sequence of the complementary strand?**

a) 5′ – TCAGCTGCT – 3′ b) 5′ – TCGTCGACT – 3′

c) 5′ – AGTCGACGA - 3′ d) Too little information to predict

19. **In Eukaryotes, the DNA replication occurs during**
    a) S phase
    b) G1 phase
    c) G2 phase
    d) All phases

20. **DNA ligase**
    a) Unwinds the helical DNA by breaking the hydrogen bonds between complementary bases
    b) Adds DNA nucleotides to the RNA primer
    c) Links the DNA fragments of the lagging strand together
    d) Synthesizes a short RNA primer at the beginning of each origin of replication

21. **The 5' end of the DNA is the one with**
    a) The terminal phosphate group on the 5' carbon of the deoxyribose
    b) A terminal hydroxyl (OH) group on the deoxyribose of the 3' carbon of the deoxyribose
    c) A terminal hydroxyl (OH) group on the deoxyribose of the 5' carbon of the deoxyribose
    d) None of the above

22. ________ **is one correct reason for why Okazaki fragments are created during lagging strand DNA synthesis**
    a) DNA is only polymerized in the 3' to 5' direction.
    b) DNA polymeraserequires RNA priming
    c) DNA is only polymerized in the 5' to 3' direction
    d) DNA polymerase requires a 5' -0H on the growing polymer

23. **At the completion of DNA replication, each newly synthesized DNA strand is**
    a) Identical in sequence to the strand opposite which it was synthesized
    b) Complementary in sequence to the strand opposite which it was synthesized
    c) A hybrid stand consisting both DNA and RNA strand
    d) Oriented in the same direction as the strand which it was synthesized

24. **Transcription, the synthesis of RNA using DNA as template, is required for**

a) Gene expression
b) DNA replication
c) Translation
d) All the above

25. **Transcription is asymmetric i.e.,**

a) both DNA strands are used as template
b) only one of the strand of DNA is used as template
c) combinations of different parts of DNA is used as template
d) all the above

26. **The nascent RNA strand synthesized as continuous strand during transcription and it is analogous to**

a) Lagging strand
b) Leading strand
c) Antiparallel strands
d) All the above

27. **The nucleotide immediately preceding the transcription unit on the DNA strand is defined as position**

a) +1
b) 0
c) -1
d) 5′ UTR

28. **RNA polymerases**

a) Require RNA primers
b) Do not require primer and initiate strand synthesis de novo
c) Do not proofread their transcripts
d) Both b and c

29. **Transcription factors recognize and binds to**

a) Silencer
b) Enhancer
c) Promoter
d) All the above

30. **Roger Kornberg got Noble prize for the year 2006 for his work on**

a) RNA Polymerase
b) Histone
c) DNA Polymerase
d) Ribosome

31. **Which of the following features would you NOT expect to find in heterogeneous nuclear RNA (hnRNA)?**
    a) Intron
    b) Polycistronic coding
    c) Polyadenylation at 3'-end
    d) 5-′ Cap structure

32. **The non-coding DNA that occurs within eukaryotic gene is referred to as**
    a) Spacer DNA
    b) Promoter
    c) Terminator
    d) Intron DNA

33. **Which of the following statements about introns is incorrect? Introns are**
    a) Found in most eukaryotic genes and not translated during protein synthesis
    b) Removed during RNA processing
    c) Normally not transcribed
    d) Responsible for the fact that the most eukaryotic mRNAs are much shorter than the genes from which they are derived.

34. **DNA is double stranded. One strand is called the coding strand and the other the non-coding strand. The non-coding strand is used as the template to make mRNA. The relationship between the base sequence of the coding strand and the base sequence of the mRNA (ignoring the fact that mRNA will contain uracil instead of thymine) is**
    a) Complementary
    b) Identical
    c) Anti-parallel
    d) None of the above

35. **The RNA strand synthesized during transcription elongates until**
    a) The entire chromosome has been copied into RNA
    b) The RNA polymerase runs into the next gene
    c) An intron encountered on the DNA template strand
    d) A specific termination sequence is reached on the DNA template strand

36. **Transcriptional regulation is controlled by**
    a) Methylation of DNA
    b) Number of genes in the genome
    c) Poly A tail
    d) Introns

37. **ESTs are obtained through**
    a) Genomic DNA library b) DNA library
    c) RT-PCR d) All the above

38. **The only one start codon is ----------**
    a) GAT b) ATG
    c) TAG d) GTA

39. **In prokaryotes, the matured mRNA is**
    a) Identical to the initial mRNA
    b) Shorter than the initial mRNA
    c) Longer than the initial mRNA
    d) None of the above

40. **In eukaryotes, translation occurs in**
    a) Cytoplasm
    b) Nucleus
    c) Cellular compartment
    d) All the above

41. **In eukaryotes, nascent mRNA is usually**
    a) Polycistronic b) Monocistronic
    c) Dicistronic d) Tricistronic

42. **The templates used for protein synthesis is**
    a) Genic DNA b) MRNA
    c) RRNA d) TRNA

43. **A sequence of translatable codons are referred as**
    a) ORF b) RBS
    c) Promoter d) Enhancer

44. **The mRNA that are stable for hours or even days are synthesized in**
    a) Eggs b) Muscles
    c) Bones d) All the above

45. **The role of modification of 5′ end of mRNA with 7-methylguanosine cap is**
    a) Ribosome binding b) Aligning p site
    c) Aligning a site d) mRNA stability

46. **The maximum portion (60-65%) of total RNA consists of**
    a) mRNA b) tRNA
    c) rRNA d) hnRNA

47. **Which site of the ribosomes binds with incoming charged tRNAs during elongation?**
    a) A (aminoacyl) site b) P (peptidyl tRNA) site
    c) GTPase domain d) Peptidyl transferase domain

48. **In eukaryotes, the small subunit of ribosome along with cap binding proteins (CBP) (eIF-4F) binds at**
    a) Shine-Dalgarno sequence
    b) Kozak sequence (ACCAUGG)
    c) Promoter
    d) Enhancer

49. **Examples of programmed misreading, where the normal interpretation of the sequence of codons is suppressed is**
    a) Read through
    b) Selenocystein insertion (at termination codon- UGA; called as Selenocystein insertion sequence)
    c) Frameshifting
    d) All the above

50. **A protein composed of 300 amino acids would be encoded by an mRNA of**
    a) 300 nucleotides b) 400 nucleotides
    c) 900 nucleotides d) 400 nucleotides

51. **The three types of RNA responsible for forming part of the structure upon which protein synthesis occurs, dictating the order of amino acids in the growing protein chain and transporting amino acids to their site of protein synthesis, respectively are**

a) mRNA, rRNA, tRNA b) tRNA, mRNA, rRNA
c) rRNA , mRNA, tRNA d) rRNA, tRNA, mRNA

52. **Anticodon is a part of the tRNA molecule that is responsible for specific attachment to the**

a) mRNA b) Amino acid
c) Ribosomal complex d) rRNA

53. **Which of the following sentence is incorrect**

a) Eukaryotic nuclear mRNAs are monocistronic
b) mRNA that carry the information for more than one type of protein is not found eukaryotic cells
c) A bacterial mRNA can carry the information for more than one type of protein
d) All the above

54. **In prokaryotes, many ribosomes translate an mRNA simultaneously, forming a structure called**

a) Polysome b) Polycistron
c) 30S initiation complex d) All the above

55. **In Eukaryotes, to locate the initiator codon, the small ribosomal subunit has to first locate**

a) The 5′-methyl guanosine cap
b) Ribosome binding site
c) Shine-Dalgarno sequence
d) All the above

56. **The process of *de novo* organ formation referred as**

a) Organogenesis b) Morphogenesis
c) Both a and b d) Rhizogenesis

57. **Production of true-to-type plants using tissue culture protocol is termed as**
    a) Micrografting b) Macropropagation
    c) Macrografting d) Micrografting

58. **Totipotency is not found in**
    a) Terminal leaf
    b) Terminal bud
    c) Terminally differentiated structures such as sieve tubes/ tracheids
    d) All the above

59. **Adventitious *de novo* formation of roots is called as**
    a) Caulogenesis b) Rhizogenesis
    c) Organogenesis d) All the above

60. **Embryos formed from cells or tissues of plant body i.e., other than zygotic or seed embryos are called as**
    a) Embryoids b) Somatic embryos
    c) Both a and b d) Plant embryos

61. **Embryos formed directly from the microspores are termed as**
    a) Androgenesis b) Gynogenesis
    c) Microgenesis d) Embryogenesis

62. **Production of haploid plants from female egg nucleus or ovum is called as**
    a) Androgenesis b) Gynogenesis
    c) Microgenesis d) Embryogenesis

63. **An individual plant selected for propagation is referred as**
    a) Mother plant b) Stock plant
    c) Ortet d) All the above

64. **Undefined' part of the tissue culture medium comprises**
    a) Tender coconut water or coconut milk
    b) Fruit juices

c) Yeast extract
d) All the above

65. **Uptake of plant nutrients is influenced by**
a) Concentration of nutrients
b) pH or physiological status of tissu
c) Temperature
d) All

66. **The possible way of uptake of nutrients by plant cells in tissue culture media is via**
a) Stomata
b) Mesophyll
c) Both a and b
d) None of the above

67. **The buffer which is commonly used in plant tissue culture media is**
a) KH_2PO_4
b) Organic acids
c) TRIS
d) All the above

68. **The first cytokinins to be discovered**
a) Kinetin
b) 6-furfuryl aminopurine
c) Both a and b
d) Dihydrozeatin

69. **Who proposed the concept of the "cell theory"?**
a) Schleiden and Schwann
b) Skoog
c) Murashige
d) Karry Mullis

70. **Name the substance added to the culture medium to induce callus induction**
a) auxins
b) cytokinins
c) GA_3
d) Ethylene

71. **Which among the following techniques does not belong to plant tissue culture**
a) Micropropagation
b) Anther culture
c) Replica plating
d) Protoplast culture

72. **Virus free plants can be produced by**
    a) Meristem tip culture b) Callus culture
    c) Somatic embryogenesis d) Suspension culture

73. **The first commercially released GM crop in India is**
    a) Cotton expressing *cryIAb* gene
    b) Brinjal expressing *cry1Ab* gene
    c) Corn expressing *cryIAb* gene
    d) Cotton expressing *cryIAc* gene

74. **The term somaclonal variation was coined by**
    a) Murashige and Skoog b) Karp and Maddock
    c) Gamborg and Phillips d) Larkin and Scowcroft

75. **During *in vitro* multiplication of plant, browning can be controlled by**
    a) Nurse culture
    b) Osmotic treatment of explant
    c) Addition of antioxidant in the medium
    d) Increasing agar concentration

76. **Cybrids are obtained by**
    a) Crossing a male sterile CMS line with a new line which is going to be converted into a CMS line
    b) By fusing cytoplasm of CMS line with normal protoplasts of the line into which CMS is to be transferred
    c) Both a and b
    d) None of the above

77. **Abscissic acid controls**
    a) Shoot elongation
    b) cell elongation and cell wall formation
    c) leaf fall and dormancy
    d) cell division

78. **Embryogenesis in plant tissue culture is induced by**
   a) Ethylene b) Abscissic acid
   c) Gibberellic acid d) Auxin

79. **Growth factors, amino acids and vitamins are sterilized by**
   a) Filter sterilization b) Chemical sterilization
   c) Steam sterilization d) Air sterilization

80. **An anti-malarial drug "Quinine" is a secondary metabolite produced by the plant**
   a) Papaver somniferum b) Jasminum spp
   c) Cinchona officinalis d) Vanilla spp

81. **Haploid production through anther culture is referred as**
   a) Sporogenesis b) Androgenesis
   c) Gynogenesis d) Organogenesis

82. **Haploids can be diploidized to produce homozygous plants by**
   a) Gibberellic acid b) Auxins
   c) Cytokinins d) Colchicine

83. **FlavrSavr® is a genetically engineered tomato showing**
   a) Delayed ripening b) Resistance against fruit borer
   c) Increased aroma d) Increased lycopene

84. **Genes present in the T-DNA of *Agrobacterium tumefaciens* are called as**
   a) Transfer genes b) Phytooncogenes
   c) Pseudogenes d) Polygenes

85. **Vitamin A rich rice will have over accumulation of**
   a) Alpha carotene b) Beta carotene
   c) Ascorbic acid d) Xanthoxin

86. **Sciophytes are**
   a) Sun loving b) Shade loving
   c) Winter crops d) Water loving

87. **Cofactor of the enzyme Urease in higher plants**
    a) Cobalt b) Nickle
    c) Zinc d) Molybdenum

88. **Group of plants produced from a single plant through asexual reproduction**
    a) Somaclone b) Tillers
    c) Clone d) Adventitious shoots

89. **Antinutritional factor present in *Brassica* spp.,**
    a) Erucic acid b) Phytic acid
    c) Rypsin inhibitor d) Haemagglutinin

90. **Generation of homozygous doubled haploids from anthers of F1 hybrids is called as**
    a) Anther culture b) Chromosome doubling
    c) Hybrid sorting d) Micropropagation

91. **Conversion of less valuable compounds into more useful compounds through cell culture is called as**
    a) Biotransformation b) Genetic engineering
    c) Secondary metabolite d) Bioconversion

92. **Chemical used in the production of artificial seeds**
    a) Agarose b) Calcium alginate
    c) Gelrite d) Acrylamide

93. **Organelle transfer is possible through**
    a) Cybrid development b) Particle gun bombardment
    c) Cell fusion d) All the above

94. **Cytoplasmic genes are called as**
    a) Epigenes b) Plasmon
    c) Karyons d) Pseudogenes

95. **--------- is a mass of undifferentiated cells**
    a) Primordia b) Somatic embryo
    c) Callus d) SAM

96. **Protoplast was discovered by — — — — — —**

a) Mendel
b) Purkinje
c) Darwin
d) Morgan

97. **First successful transgenic fruit plant is produced in**

a) Apple
b) Pear
c) Plum
d) Peach

98. **Molecular cloning is an _____ technique.**

a) In vivo
b) In vitro
c) Both a and b
d) None of the above

99. **The restriction enzyme which is used widely in molecular cloning experiments are**

a) Type I
b) Type II
c) Type III
d) All the above

100. **Source of DNA for cloning experiments is**

a) Genomic DNA
b) cDNA
c) PCR product or previously isolated clone or chemically synthesized oligonucleotides
d) All the above

101. **Random cleavage of DNA can be achieved by**

a) Mechanical shearing (sonication, vortexing)
b) Chemical treatment (acid/alkali hydrolysis)
c) Non-specific endonuclease
d) All the above

102. **Precise, sequence specific cutting of DNA by restriction enzymes usually produce**

a) Blunt end
b) Sticky end
c) Both a and b
d) None of the above

103. **The 5′ or 3′ end of DNA can be modified by**

a) End filling and trimming with T4 DNA polymerase
b) Partial end filling with Klenow DNA polymerase

c) Addition of adaptors/linkers followed by restriction enzyme digest
d) All the above

104. Double stranded oligonucleotides which contain restriction site used in DNA end modification are

a) Linkers
b) Adoptors (predigested linkers with restriction enzymes)
c) Oligonucleotides
d) Primers

105. **Predicting the orientation of insert DNA in the vector is call**

a) Directional cloning (eg. use of pair of restriction enzymes)
b) Bidirectional cloning
c) Both a and b
d) None of the above

106. **Addition of nucleotides to the 3′ end of Blunt or ragged donor DNA by terminal dexoynucleotidyl transferase is termed as**

a) homopolymer tailing b) poly-A insertion
c) both a and b d) none of the above

107. **An ideal cloning vector displays which one of the following property?**

a) Episomal (not integrated into host genome)
b) Replicate autonomously
c) Allow selection of recombinant from non-recombinants
d) All the above

108. **An important initial step in development of naturally occurring plasmids and Bacteriophages as cloning vector is**

a) Construction of multiple cloning site
b) Retaining selectable or visible marker gene
c) Removal of unwanted DNA
d) All the above

109. **Usually the plasmids are developed with**
   a) Col E1 replicon
   b) Bacteriophage M13 origin of replication
   c) Both a and b
   d) None of the above

110. **Selection of recombinant DNA can be made easy by using**
   a) Insertion vectors b) Replacement vectors
   c) Both a and b d) None of the above

111. **Assembly of recombinant vector into Bacteriophage ë heads is called as**
   a) In vitro packaging b) Wrapping up
   c) Both a and b d) None of the above

112. **Usually the scorable or selectable markers used in vectors are**
   a) Dominant b) Recessive
   c) Epistatic expression d) All the above

113. **Blue white selection strategy used to select recombinant cells is based on**
   a) β-Galactosidase b) β-Glucosidase
   c) Pepetidogalactosidase d) Galactose

114. **Recombinant λ insertion vectors are usually selected by disruption of cI gene which generates ____ plaques**
   a) Clear b) Turbid
   c) Both a and b d) None of the above

115. **λ and cosmid vectors form infection particles if the recombinant genome is ___ % of the wild type genome size**
   a) 100 b) 75-100
   c) 105 d) 75

116. **Chromosome specific genomic library can be generated by chromosome separation by**
   a) fluorescence activated chromosome sorting (FACS)
   b) monochromosomal somatic cell hybrids

c) chromosome microdissection

d) all the above

117. **Which of the following pairs are anome.**

a) Ribose and Ribulose

b) Glucose and mannose

c) Pyruvate and lactace

d) a-D-glucose and b-D-glucose

118. **Hybridoma technology is used in the production of**

a) mRNA b) Monoclonal serum

c) Monoclanal antibodies d) Polyclonal antibodies

119. **For mitochondrial structure and faction mt-DNA specifies**

a) 10% b) 20%

c) 30% d) 40%

120. **Which of the following reagent is useful for visualizing DNA**

a) Uracil b) DNA polymerase

c) Diphenylamine d) Ethidium bromide

121. **Homologous pairs line up along equatorial plane during**

a) Anaphase II b) Methaphase I

c) Telophase d) Interphase

122. **Most plant cells are surrounded by a rigid cell wall made primarily of**

a) Monosaccharides b) Pectin

c) Chitins d) polysaccharides

123. **The organelle that function in the breakdown of cell and tissues in the**

a) Episomes b) Mitochondrion

c) Centrosomes d) Lysosomes

124. **Endoplasmic reticulam**
   a) Is found only in animal
   b) Is a system of membrane bound channels
   c) Is called rough if mitochondria is attached to it
   d) Is a site of ATP production

125. **9+2 describes the basic structure of which of the following one**
   a) Chromosome
   b) Basal body
   c) Chloroplast
   d) Flagellum

126. **The function of nucleus includes**
   a) Cellular respiration
   b) Synthesis of protein
   c) Housing the hereditary information
   d) Synthesis of carbohydrates

Answers

1 a 2 d 3 d 4 d 5 a 6 b 7 a 8 a 9 d 10 a
11 a 12 a 13 a 14 c 15 d 16 c 17 b 18 b 19 a 20 c
21 a 22 c 23 b 24 d 25 b 26 b 27 c 28 d 29 d 30 a
31 b 32 d 33 c 34 a 35 d 36 a 37 b 38 b 39 a 40 a
41 b 42 b 43 a 44 a 45 a 46 c 47 a 48 b 49 d 50 c
51 c 52 a 53 b 54 a 55 c 56 c 57 d 58 c 59 b 60 c
61 a 62 b 63 d 64 d 65 d 66 a 67 d 68 c 69 a 70 a
71 c 72 a 73 a 74 d 75 c 76 b 77 c 78 b 79 a 80 c
81 b 82 d 83 a 84 b 85 b 86 b 87 b 88 c 89 a 90 c
91 a 92 b 93 a 94 b 95 c 96 b 97 b 98 c 99 b 100 d
101 d 102 c 103 d 104 a 105 a 106 a 107 d 108 d 109 a 110 c
111 a 112 a 113 a 114 a 115 b 116 d 121 b 122 d 123 d 124 b
125 d 126 c

□□□

Chapter 8

Dairy Science and Technology

1. **White revolution is related to**
   a) Milk b) Cereal
   c) Egg d) None of these

2. **What is the principal carbohydrate in the milks of all mammals?**
   a) Lactose b) Glucose
   c) Sucrose d) Fructose

3. **The father of white revolution is**
   a) V.Kurien b) L .kurien
   c) S. Swaminathan d) N.E Borlaug

4. **Off-flavor of milk that may be described as tasteless and the flavor of normal milk is lacking.**
   a) High acid b) Rancid
   c) Oxidized d) Flat/watery

5. **Lactose can be prepared from**
   a) Milk b) Milk sugar
   c) Curd d) Paneer

6. **Percentage of water in buffalo milk is: -**
   a) 65-67 % b) 70-75 %
   c) 80-85 % d) 87-90 %

7. **Lactometer is the instrument used for**
   a) Specific gravity of milk b) Density of milk
   c) SNF of milk d) TSS of milk

8. **Hormone responsible for milk synthesis is**

a) Elastin b) Prolactin

c) Colostrum d) Cobalamin

9. **Flavor detected in milk by consumption of strong feeds or weeds or milk from cows in late stages of lactation.**

a) Feed b) Salty

c) Bitter d) Rancid

10. **Lactic acid content in fresh milk is**

a) 0.16-0.21% b) 0.65-0.78%

c) 0.14-0.15% d) 0.25-0.36%

11. **Soft fats in milk fat are:-**

a) Lauric & Stearic b) Capric & Lauric

c) Oleic & Butyric d) Oleic & Lauric

12. **Fat supplies how much time more energy than that of carbohydrates**

a) 2.28 b) 2.25

c) 2.26 d) 2.68

13. **Off-flavors are imparted to milk when cows eat obnoxious weeds. Flavor can be recognized by distinctive odor and taste suggested by its name.**

a) Rancid b) Garlic/onion

c) Foreign d) Flat/watery

14. **Water soluble vitamins**

a) B b) C

c) Both A&B d) A

15. **Fat content in double toned milk is**

a) 5.2% b) 1.5%

c) 1.8% d) 2.6%

16. **Principal protein in milk is:-**
   a) Albumin b) Lactalbumin
   c) Casein d) Lactoglobulin

17. **Colour of buffalo milk is**
   a) White b) Yellow
   c) Dark yellow d) Both A & B

18. **Whey is the by-product in the manufacture of:-**
   a) Skimmed milk b) Butter
   c) Cheese d) Yogurt

19. **Better quality khoya is prepared from**
   a) Buffalo milk b) Cow milk
   c) Goat milk d) Both A & B

20. **Serious off-flavor to milk caused by fly spray, paint, kerosene or creosote. Substances enter milk as direct contaminants or in vapor form.**
   a) Rancid b) Foreign
   c) Feed d) Malty

21. **Fat soluble vitamins include**
   a) A & D b) E & K
   c) B & C d) Both A & B

22. **The Anand pattern started with**
   a) One village b) Four villages
   c) Only two villages d) Three villages

23. **Example of soft cheese is:-**
   a) Cheddar b) Swiss
   c) Brick d) Cottage

24. **Solid not fat (SNF) in cow milk**
   a) 9.2% b) 5.6%
   c) 8.5% d) 8.9%

25. **Off-flavor is quite pungent in advanced stages and is recognized by its "papery" or cardboardsensation initially and a tallow odor in more advanced stages.**

a) Malty b) Foreign
c) Salty d) Oxidized

26. **Glucose fructose cellulose hemi cellulose are**

a) Vitamin b) Fat
c) Milk sugar d) Carbohydrate

27. **Which one of the following is not fermented beverages?**

a) Kefir b) Leban
c) Buttermilk d) Kaumiss

28. **Inflammation of udder is known as**

a) Mastitis b) Anthrax
c) Cow pox d) Rinderpest

29. **This flavor resembles the flavor of stale fat and has a soapy taste and a goaty, unclean odor as a result of breaking down of fats in milk.**

a) Salty b) Garlic/onion
c) Feed d) Rancid

30. **The sugar present in milk is**

a) Fructose b) Lactose
c) Maltose d) Sucrose

31. **How many indigenous enzymes have been reported in bovine milk?**

a) 30 b) 60
c) 50 d) 40

32. **Fat globule in the milk is present in the form of**

a) Emulsion b) Colloidal
c) Suspension d) None of these

33. **This off-flavor is seldom found except in pasteurized milk that has been stored too long or stored at a slightly high refrigerator temperature.**

a) Foreign
b) Unclean
c) High acid
d) Feed

34. **Â- carotene is the precursor of**

a) Vitamin D
b) Vitamin A
c) Vitamin E
d) Vitamin K

35. **Lactose has water solubility of**

a) 100%
b) 50%
c) 18%
d) 25%

36. **Ash content of human milk is**

a) 0.6%
b) 0.2%
c) 0.3%
d) 05%

37. **Destruction of which enzyme is used as an index of super-HTST pasteurization?**

a) Catalase
b) Lipase
c) Lactoperoxidase
d) All of the above

38. **Exposing of milk at 61-63 C for minutes is**

a) Low temperature long time
b) High temperature short time
c) Ultra high temperature
d) High temperature long time

39. **Milk not properly cooled may encounter this off flavor and is caused by improperly cleaned milking machines or equipment.**

a) Malty
b) Foreign
c) Feed
d) Rancid

40. **Solid not fat (SNF) content in goat milk**

a) 9.0%
b) 8.5%
c) 7.6%
d) 8.4%

41. **Casein from of total milk protein**
   a) 85-90% b) 90-92%
   c) 60-75% d) 80-83%

42. **Enzyme coagulated milk product is**
   a) Paneer b) Cheese
   c) Ghee d) Curd

43. **Rennet belongs to:-**
   a) Lipases b) Catalase
   c) Proteinases d) Phosphatases

44. **Cow milk is generally richer than buffalo milk in**
   a) â-carotene b) á-carotene
   c) Both A& B d) None of these

45. **Average pH of milk**
   a) 6.8 b) 6.6
   c) 6.5 d) 6.7

46. **Gerber method is used for testing of**
   a) Fat in milk b) SNF in milk
   c) TSS in milk d) All of the above

47. **Temperature used in UHT treatment is:-**
   a) 90-100 °C b) 100-120 °C
   c) 120-125 °C d) 130-140 °C

48. **Hormone for milk secretion**
   a) Prolactin b) Albumin
   c) Perosis d) None of these

49. **Removing offending feeds from cows four hours before milking, can reduce the risk of getting this off-flavor milk.**
   a) Bitter b) Acid
   c) Feed d) Malty

50. **Methylene blue reduction test is conducted to know the quality of**

a) Pasteurized milk b) Toned milk
c) Skim milk d) Fresh milk

51. **Cream line dairy organization brand name is**

a) Amul b) Anand
c) Jersey d) Gir

52. **Lactose is disaccharide containing: -**

a) Glucose & Fructose b) Glucose & Galactose
c) Glucose & Glucose d) Glucose & Maltose

53. **Freshly drawn milk has a pH value of**

a) 6.8-6.9 b) 5.9-6.3
c) 6.7-7.2 d) 6.5-6.7

54. **The human milk contain SNF …….. amount**

a) Maximum b) Minimum
c) Equal d) None of these

55. **A CMT test appearance with a strong gel formation that tends to adhere to paddle and forms a distinct central peak would have a Leukocyte count/ml of__________.**

a) Below 200,000 b) 150,000-500,000
c) 800,000-5,000,000 d) Over 5,000,000

56. **Percentage of fat is minimum in the milk of**

a) Brawn swiss b) Holstein Friesian
c) Frieswal d) Jersey

57. **NDDB is situated at**

a) Gujarat b) Hissar
c) Karnal d) Bangalore

58. **Milkman is a brand name of**

a) Britannia organization b) Amul organization
c) Jersey organization d) All of the above

59. **CaCl2 is added at the rate of:-**

a) 0.5 % b) 0.8 %

c) 0.02 % d) 0.08 %

60. **NDDB was established in which five-year plan**

a) 3rd b) 4th

c) 5th d) 2nd

61. **Dairy product resulting from the addition of a creaming mixture (dressing) that contains not less than 4 percent milk fat and not more than 80 percent moisture?**

a) Sorbet b) Sherbet

c) Yogurt d) Cottage cheese

62. **Milk is poor source of**

a) Vitamin A b) Iron

c) Vitamin C d) both B & C

63. **Which one is used as an emulsifying agent in process cheese blend?**

a) Paprika b) Pectin

c) Glycerides d) Whay Powder

64. **Pasteurization kills**

a) Bacteria spores alive b) Virus

c) Both A&B d) Minerals

65. **Percentage of total lactose is maximum in the milk of cattle breed known as**

a) Ongole b) Sahiwal

c) Rathi d) Tharparker

66. **The boiling point of milk is**

a) 100.25 C b) 100.8 C

c) 102.5 C d) 100.17 C

67. **Dairy Rations for lactating cows should be calculated based on:**

a) Body Size b) Milk Production
c) Stage of lactation d) All of these

68. **The fat content of cream should not be less than**

a) 25% b) 35%
c) 68% d) 53%

69. **Coagulated milk product is**

a) Curd b) Paneer
c) Ghee d) All of the above

70. **Milk is an emulsion: -**

a) Oil-in-Water b) Water-in-Oil
c) Oil-in-Oil d) Oil-in-Starch

71. **Ergosterol is a precursor of**

a) Vitamin E b) Vitamin D
c) Vitamin A d) None of these

72. **Which of the following would be classified as a "Hard Cheese"?**

a) Brick b) Montery Jack
c) Cheddar d) Brie

73. **Fat separated milk should not be more than**

a) 0.4% b) 0.26%
c) 0.5% d) 0.36%

74. **Glucose + fructose =**

a) Sucrose b) Fructose
c) Maltose d) Galactose

75. **Concept of pasteurization was given by**

a) Louis Pasteur b) William
c) S. gaud d) None of these

76. **Normal bovine milk contains protein about**
    a) 7.5% b) 5.5%
    c) 3.5% d) 9.5%

77. **Fat content in double toned milk**
    a) 1.25 b) 1.69
    c) 2.86 d) 1.55

78. **As it comes from a cow, the solids portion of milk contains approximately 3.7 percent fat and ____ percent solids-not-fat.**
    a) 3% b) 6%
    c) 9% d) 12%

79. **Glucose + glucose =**
    a) Sucrose b) Fructose
    c) Maltose d) Galactose

80. **Bacteria responsible for preparation of dahi is**
    a) Streptococcus lactic b) Streptococcuslactic
    c) Streptococcus d) None of these

81. **Which enzyme is tested for cream pasteurization?**
    a) Plasmin b) Phosphates
    c) Catalase d) peroxidase

82. **Khoya contains milk fat not less than**
    a) 20% b) 25%
    c) 48% d) 30%

83. **A good quality curd has acidity less than**
    a) 0.8% b) 0.56%
    c) 0.258% d) 0.7 %

84. **The aim of pasteurization milk is to:-**
    a) Improve flavor
    b) Kill disease producing organisms
    c) Improve color
    d) None of the above

85. **Buffer value of milk is due to**

a) Fat b) Protein

c) Minerals d) carbohydrate

86. **People with high blood pressure or edema are advised to take:-**

a) Multivitamin Mineral Milk

b) Low Sodium Milk

c) Sterile Milk

d) Low Lactose Milk

87. **Oxidation of fat is known as**

a) Beta oxidation b) Gama oxidation

c) Alpha oxidation d) All of the above

88. **Sugar content of milk in cattle**

a) 2.6-2.9% b) 4.5-4.8%

c) 5.6-6.8% d) 4.0-4.5%

89. **Evaporated milk is concentrated to approximately what times the solid of normal whole milk?**

a) 2.25 times b) 6.25 times

c) 7 times d) 8.5 times

90. **Percentage of total solids is maximum in the milk**

a) Brawn swiss b) Holstein Friesian

c) Frieswal d) Jersey

91. **Most successful example of cooperative in milk is**

a) AMUL b) ANAND

c) Both A & B d) None of these

92. **Milk contains 87 % water and the rest is solids and fats. Which of the following is not included as a milk solid?**

a) Protein b) Water

c) Carbohydrate d) Milkfat

93. **Operation flood – 1 program was inaugurated**
   a) 1970 b) 1980
   c) 1966 d) 1975

94. **The freezing point of cow and buffalo milk varies from**
   a) -0.258 C to -0.289 C b) -0.512 C to -0.572 C
   c) -0.487 C to -0.498 C d) -0.187 C to -0.235 C

95. **10-15 % more milk is produced with which growth hormones, if injected to lactating cows?**
   a) Auxin b) Bovine Growth Hormone
   c) Ethylene d) None of the above

96. **Operation flood is related to**
   a) Milk b) Cereal
   c) Fertilizer d) All of the above

97. **The process to increase in volume caused by whipping air into the ice cream mix during freezing is called?**
   a) Homogenization b) Aging
   c) Overrun d) Hardening

98. **The water content of milk is**
   a) 85-90% b) 90-98%
   c) 80-90% d) 70-80%

99. **The state most progressive in dairy farming is**
   a) Panjab b) Haryana
   c) Gujrat d) All of the above

100. **The part of the mammary gland where milk produced is called:**
   a) Capillary b) Teat Cistern
   c) Chine d) Alveolus

101. **The milk stone is removed by**
   a) Basic b) Acid
   c) Alkaline solution d) None of these

102. **The exotic breed that has maximum lactose in milk**
   a) Brawn swiss b) Holstein Friesian
   c) Frieswal d) Jersey

103. **Semisolid ice cream is placed in a hardening room at a temperature of about?**
   a) -15 °C b) -20 °C
   c) -34 °C d) -44 °C

104. **The normal content of protein in cow milk**
   a) 2.6% b) 5.6%
   c) 3.5% d) 4.6%

105. **The non-reducing carbohydrate is**
   a) Lactose b) Glucose
   c) Sucrose d) None of the above

106. **The *"eyes"* in swiss cheese are formed by the growth of:-**
   a) Leuconostoc b) Propionibacterium
   c) Streptococcus d) Lactobacillus

107. **The protein percentage in cow milk is usually in the ranges of**
   a) 3.3-4.3 b) 3.6-3.9
   c) 5.4-5.9 d) 2.89-3.19

108. **The exotic breed that has minimum milk fat percentage is**
   a) Bhadawari b) Holstein Friesian
   c) Frieswal d) Jersey

109. **The Primary Milk carbohydrate is:**
   a) Leucine b) Sucrose
   c) Arginine d) Lactose

110. **The optimum temperature for making curd**
   a) 22 C b) 30 C
   c) 58 C d) 43 C

111. **The most variable constituent of milk is**

a) Fat
b) Protein
c) Minerals
d) Carbohydrate

112. **Yogurt contains mixed lactic acid culture containing: -**

a) Lactobacillus bulgaricus and Streptococcus thermophilus
b) Lactobacillus bulgaricus and Propionibacterium
c) Lactobacillus bulgaricus and Leuconostoc
d) None of the above

113. **The SNF value of tonned milk is**

a) 8.5%
b) 7.6%
c) 8.0%
d) 9.2 %

114. **The heritability of the traits for milk production is:**

a) 15%
b) 25%
c) 55%
d) 100%

115. **The most suitable temperature for separation of cream by centrifugal cream separator is**

a) 50 C
b) 40 C
c) 68 C
d) 72 C

116. **The normal titratable acidity in fresh cow milk is**

a) 0.12%
b) 0.35%
c) 0.5%
d) 0.15%

117. **The part of the mammary gland where milk is produced is called:**

a) Capillary
b) Teat Cistern
c) Chine
d) Alveolus

118. **The optimum temperature for the growth of mesophilic bacteria**

a) 30-32 C
b) 35-40 C
c) 55-60 C
d) 25-29 C

119. **Oxidized flavor development is catalyzed by**

a) Copper b) Iron
c) Both A & B d) None of these

120. **First milk of parturition cattle is known as**

a) Cobalamin b) glycogen
c) Colostrum d) None of the above

121. **The melting point of milk fat varies normally between:-**

a) 32-36 °C b) 40-45 °C
c) 20-25 °C d) 26-30 °C

122. **Protein present in the milk is in the form of**

a) Colloidal b) Emulsion
c) Fat d) All of the above

123. **The specific gravity of milk fat at 21°C is?**

a) 0.70 b) 0.82
c) 0.93 d) 0.98

124. **Operation flood -2 was inaugurated on**

a) 1978 b) 1980
c) 1850 d) 1995

125. **Milk is a rich source of**

a) Ca b) p
c) Vitamin A d) All of the above

126. **The iodine number measures the amount of:-**

a) Free Fatty Acids
b) Saturated Fatty Acids
c) Chain Length of Fatty Acids
d) Unsaturated Glycerides

127. **The operation flood project conceived and formulated in**

a) 4th FYP b) 5th FYP
c) 3rd FYP d) 2^{nd}FYP

128. **Milk fat differs from other common fats in having a larger percentage of:-**
   a) Free Fatty Acids b) Saturated Fatty Acids
   c) Volatile Fatty Acids d) None of the above

129. **The most potent source of bacterial contamination in milk is**
   a) Utensils b) Colostrum
   c) Hypochlorite d) All of the above

130. **Saponification number of butter fat is:-**
   a) 190 b) 195
   c) 210 d) 231

131. **The milk sugar is**
   a) Disaccharide b) Monosaccharide
   c) Trisaccharide d) None of the above

132. **Energy value of milk having 3.5% fat according to Sharp's Formula will be?**
   a) 60 b) 60.5
   c) 62.5 d) 65.5

133. **The moisture content of paneer is not less than**
   a) 70% b) 80%
   c) 75% d) 65%

134. **The most common and inexpensive chemical for sanitizing the dairy utensil is**
   a) Hypochlorite b) Chlorites
   c) Nitrite d) None of these

135. **Lactose reduces copper salt to?**
   a) Cupric Oxide b) Cuprous Oxide
   c) Copper Oxide d) None of the above

136. **Yellow colour of cow milk is due to the presence of**
   a) Carotene b) Casein
   c) Both A & B d) Lactose

137. **Yellow color of milk fat is due to presence of?**
   a) Vitamin D b) Carotenoids
   c) Calcium d) Folic Acid

138. **UHT milk is heated to a temperature**
   a) 145 C b) 150 C
   c) 160 C d) 158 C

139. **Tests for proper pasteurization are based on the activity of which enzyme?**
   a) Lactase b) Diastase
   c) Phosphatase d) Catalase

140. **The specific gravity of cow milk ranges between**
   a) 1.028-1.032 b) 2.036-2.058
   c) 1.010-1.026 d) 0.0-1.023

141. **Curd tension is**
   a) <milk b) >milk
   c) =milk d) None of these

142. **Starch is split by which enzyme?**
   a) Lactase b) Diastase
   c) Phosphatase d) Catalase

143. **The pH of fresh milk is-**
   a) 6.2-6.5 b) 6.7-6.9
   c) 6.5-6.7 d) 6.5-7.2

144. **The average freezing point of milk is**
   a) 0.05 b) 0.5%
   c) 0.26% d) -0.545 C

145. **What is freezing point of milk?**
   a) 0 °C b) -0.55 °C
   c) -1 °C d) -1.55 °C

146. **Which is not an essential amino acid**
   a) Arginine b) Glycine
   c) lysine d) All of the above

147. **Concentration contain-**
   a) CF<16%, TDN>70% b) CF<12%, TDN>50%
   c) CF<18%, TDN>60% d) None of these

148. **What is the average boiling point of milk?**
   a) 100-101 °C b) 105-110 °C
   c) 115-117 °C d) 102-105 °C

149. **Amount of water required for 1 kg dry matter feed by –**
   a) 3-5 kg b) 2-3 kg
   c) 5-7 kg d) 8-10 kg

150. **What is the average specific gravity of normal whole milk at 16 °C?**
   a) 0.903 b) 1.032
   c) 1.582 d) 2.032

151. **Buffaloes rolling or floundering in mud or water is called –**
   a) Wallowing b) Bellowing
   c) Weaning d) All of these

152. **Which is the pre-dominating organism in dirty utensils of milk?**
   a) Lactobacillus bulgaricus
   b) Propionibacterium
   c) Streptococcus lactic
   d) All of the above

153. **Project directorate on cattle, Meerut came into existence in –**
   a) 1974 b) 1985
   c) 1987 d) 1990

154. **NDDB at Anand Gujarat was established in**
   a) 1952 b) 1963
   c) 1966 d) 1970

155. **Energy value of a food is measured in terms of?**
a) Carbohydrates b) Fats
c) Proteins d) Calories

156. **A pregnant cow does not come in heat because –**
a) Corpus luteum is present
b) Corpus luteum is absent
c) Graafian follicle is present
d) None of these

157. **Legal butter must contain at least what percentage of fat?**
a) 70 % b) 80 %
c) 90 % d) 95 %

158. **1st MDF in India was started at**
a) Karnal b) Hisar
c) Gujrat d) Allahabad

159. **The high nutritive value of cheese is due to:-**
a) High mineral contents b) High protein contents
c) Taste & flavor d) All of the above

160. **1st MDF in India was started in**
a) 1880 b) 1850
c) 1889 d) 1993

161. **Food value of ice cream depends to a large extent on its?**
a) Flavor b) Volume
c) Composition d) None of the above

162. **Fat percentage in ghee should be**
a) 80% b) 89%
c) 99% d) 92%

163. **A broiler will be ready for market in**
a) 35 days b) 50 days
c) 70 days d) 90 days

164. **Chlorine compounds have widespread acceptance in the dairy industry due to?**
 a) Non-toxicity b) High sanitizing efficiency
 c) High corrosiveness d) All of the above

165. **On dry matter basis chhaina should contain fat not less than**
 a) 30% b) 43%
 c) 50% d) 60%

166. **At what concentration chlorine sanitizing solutions are usually used in the dairyindustry?**
 a) 50-100 ppm b) 100-200 ppm
 c) 200-500 ppm d) 500-800 ppm

167. **The overrun in ice cream is**
 a) 100% b) 102%
 c) 101% d) 99%

168. **The process of raising or lowering the percent of fat in milk or cream to a desired standard is called?**
 a) Enrichment b) Fortification
 c) Standardization d) None of the above

169. **Highest heat transfer value is of**
 a) Copper b) Aluminium
 c) Stainless steel d) Carbon steel

170. **A product consisting of a mixture of milk and cream which contains not less than 10.5% milk fat is called?**
 a) Concentrated Milk b) Low Fat Milk
 c) Half-and-Half d) Eggnog

171. **In HTST pasteurization milk is heated to at least**
 a) 72.5 C b) 71.5 C
 c) 85.1 C d) 63.5 C

172. **The time temperature combination for proper pasteurization should be such that it gives a negative test for**

a) Alkaline phosphates b) Acidic phosphates
c) Carotene d) None of these

173. **Butter milk is a fluid product resulting from the manufacture of?**

a) Cheese b) Yogurt
c) Ice cream d) Butter

174. **Better quality chhena is prepared from**

a) Buffalo milk b) Goat milk
c) Cow milk d) Sheep milk

175. **Which of the following has largest particle size in milk?**

a) lactose b) Casein micelles
c) Fat globule d) Maltose

176. **Maintenance of lactation is**

a) Galactopoietic b) Gerber
c) Leucopoiesis d) None of these

177. **What is the application of any effective method or substance to a clean surface for destruction of pathogen is called?**

a) Pasteurization b) High Temperature Treatment
c) Sanitization d) Cleaning

178. **Milk sugar is present in milk in**

a) Fresh solution b) Ture solution
c) False solution d) None of these

179. **The state in India where per capita availability of milk is the highest is**

a) Punjab b) Hariyana
c) Gujrat d) Uttar Pradesh

180. **In what from Formaldehyde- preservatives used in milk exists?**
   a) Gas b) Liquid
   c) Solid d) All of the above

181. **Under infection generally reduces in milk**
   a) Lactose percent b) Casein percent
   c) Carotene percent d) Both A & B

182. **The quality of milk product that is affected by the presence of bacteriophage in milk is**
   a) Cheese b) Paneer
   c) Curd d) Ghee

183. **Mammary gland's complete unit of milk synthesis of is called**
   a) lumen b) Micelles
   c) Alveolus d) Secretary cell

184. **Colloidal suspension is the example of**
   a) Milk protein b) Milk mineral
   c) Both A & B d) None of these

185. **Fatty acids synthesized in mammary gland are**
   a) Higher chain fatty acids
   b) Unsaturated fatty acids
   c) Lower chain fatty acids
   d) Medium and lower chain fatty acids

Answers

1	a	2	a	3	a	4	d	5	a	6	d	7	a	8	b	9	c	10	c
11	c	12	b	13	b	14	c	15	b	16	c	17	a	18	c	19	a	20	b
21	d	22	c	23	d	24	c	25	d	26	d	27	c	28	a	29	d	30	b
31	b	32	a	33	b	34	b	35	c	36	b	37	c	38	a	39	a	40	a
41	d	42	b	43	c	44	a	45	b	46	a	47	d	48	a	49	c	50	a
51	c	52	b	53	d	54	a	55	d	56	b	57	a	58	a	59	c	60	a
61	d	62	d	63	c	64	a	65	b	66	d	67	d	68	a	69	b	70	a
71	b	72	c	73	c	74	a	75	a	76	c	77	d	78	c	79	c	80	a

81 d	82 a	83 d	84 b	85 b	86 b	87 a	88 d	89 a	90 d
91 a	92 b	93 a	94 b	95 b	96 a	97 c	98 c	99 b	100 d
101 b	102 a	103 c	104 c	105 c	106 b	107 a	108 b	109 d	110 a
111 a	112 a	113 a	114 b	115 b	116 d	117 d	118 a	119 c	120 c
121 a	122 a	123 c	124 a	125 d	126 d	127 a	128 c	129 a	130 d
131 a	132 b	133 a	134 a	135 b	136 c	137 b	138 a	139 c	140 a
141 a	142 b	143 c	144 d	145 b	146 b	147 c	148 a	149 a	150 b
151 a	152 c	153 c	154 c	155 d	156 a	157 b	158 d	159 b	160 c
161 c	162 c	163 c	164 b	165 c	166 c	167 a	168 c	169 c	170 c
171 b	172 a	173 d	174 c	175 c	176 a	177 c	178 b	179 a	180 a
181 d	182 a	183 c	184 c	185 c					

❑❑❑

Chapter 9

Entomology

1. **The gladiator belongs to the insect order**
   a) Phasmida b) Orthoptera
   c) Grylloblattoidea d) Mantophasmatodea

2. **The differentiation of thebody into distinct functional regions is called**
   a) Morphogenesis b) Oogenesis
   c) Tagmosis d) Blastogenesis

3. **In which type of head,the mouthparts are in continuous series with the legs?**
   a) Hypognathous b) Prognathous
   c) Ophisthognathous d) Paurometabolous

4. **The Johnston's organ is present on**
   a) Scape b) Pedicel
   c) Coxa d) Trochanter

5. **In silk worm, the antennae are**
   a) Clavate b) Capitate
   c) Pectinate d) Bipectinate

6. **Piercing and sucking mouthparts are found in**
   a) Mosquitoes b) Honey bees
   c) Thrips d) Butterflies

7. **Saltatoriallegs are found in**
   a) Cockroach b) Praying mantid
   c) Grasshopper d) Mole cricket

8. **The forewing of Coleoptera is called**
   a) Tegmina b) Elytra
   c) Hemelytra d) Halteres

9. **The larva which has no legs and is very poorly sclerotized is called**
   a) Oligopod b) Apodous
   c) Scarabaeiform d) Polypod

10. **Foregut is lined by**
    a) Intima b) Peritrophic membrane
    c) Enteric epithelium d) Taenidia

11. **Salivary glands are also called**
    a) Mandibular glands b) Maxillary glands
    c) Labial glands d) Pharyngeal glands

12. **A respiratory system in which 1 meso thoracic and 8 abdominal spiracles are functional is called**
    a) Holopneustic b) Perineustic
    c) Hemipneustic d) Oligopneusic

13. **The neurons which carry impulses from the central nervous system arecalled**
    a) Sensory b) Afferent
    c) Motor d) Interneurons

14. **The sac for storage of sperms in male insect is called**
    a) Spermatheca b) Seminal vesicle
    c) Median oviduct d) Genital chamber

15. **The amount of uric acid in excreta of insects is about**
    a) 10% b) 25%
    c) 40% d) 85%

16. **Juvenile hormone is secreted by**
    a) Neurosecretary cells b) Prothoracic glands
    c) Corpora allata d) Corpora cardiaca

17. **The complete dependence of one organism over another is called**
    a) Amensalism b) Mutualism
    c) Protocooperation d) Commensalism

18. **The theory of natural regulation of populations based on genetic feedback mechanism was proposed by**
    a) D. Pimentel b) D. Chitty
    c) A. Milne d) A.I. Nicholson

19. **Domestic quarantines were enforced in India to prevent the spread of**
    a) Woolly apple aphid b) Pink bollworm
    c) Brown planthopper d) San Jose scale

20. **The insect which entered India before the enforcement of quarantine measures is 2020**
    a) Cottony cushion scale b) Cotton whitefly
    c) Guava mealy bug d) Sugarcane pyrilla

21. **The Prevention of Food Adulteration Act, 1954 came into force in**
    a) 1954 b) 1955
    c) 1956 d) 1957

22. **The recommended trap crop for IPM of diamondback moth on cabbage is**
    a) Radish b) Carrot
    c) Indian mustard d) Cotton

23. **The recommended trap crop for IPM of tomato fruit borer is**
    a) Brinjal b) African marigold
    c) Potato d) Cucurbits

24. **Clipping off the top of rice seedlings containing immature stages of insects reduces the carry over of infestation of**
    a) Rice hispa b) Whitebacked planthopper
    c) Green leafhopper d) Rice bug

25. **Sticky bands around tree trunks provide protection against**
    a) Citrus psylla b) Mango mealy bug
    c) Mango hopper d) Apple root borer

26. **The first successful attempt to utilize biological control involved the importation of**
    a) Ladybird beetle b) Aphelinus spp.
    c) Vedalia beetle d) Apanteles spp.

27. **The All India Coordinated Research Project on Biological Control of Crop Pests and Weeds (AICRPBC) was started in**
    a) 1967 b) 1977
    c) 1980 d) 1985

28. **The Project Directorate of Biological Control (PDBC) came into being in**
    a) 1963 b) 1970
    c) 1983 d) 1993

29. **The cassava mealy bug in Africa was successfully controlled by importation and augmentation of**
    a) Cyrtorhinus lividipennis (Reuter)
    b) Cryptolaemus montrouzieriMulsant
    c) Epidinocarsis lopezi (De Santis)
    d) Aphelinus mali (Haldeman)

30. **San Jose scale has been successfully controlled in India by the importation of**
    a) Aphelinus mali (Haldeman)
    b) Encarsia perniciosi (Tower)
    c) Coccinella septempunctata Linnaeus
    d) Trichogramma chilonis Ishii

31. **The beetle, Zygogramma bicolorata Pallinter, introduced for the control of congress grass, started feeding on**
    a) Cauliflower b) Mustard
    c) Sunflower d) Soybean

32. **A strain of Trichogramma chilonis Ishii has been developed in India which is resistant to**

a) Endosulfan b) Malathion
c) Carbaryl d) Carbofuran

33. **The green lacewing is a parasitoid of**

a) Beetles b) Grasshoppers
c) Aphids d) Cockroaches

34. **The first record of an insect disease caused by a fungal pathogen is that of**

a) *Beauveria bassiana*
b) *Verticillium lecanii*
c) *Cephalosporium lecanii*
d) *Metarrhizium anisopliae*

35. ***Bacillus thuringiensis* was first isolated from diseased larvae of**

a) Silkworm b) Honey bee
c) American bollworm d) Spruce budworm

36. ***Bacillus papillae* causes milky disease in**

a) Cockchafer beetle b) Japanese beetle
c) Red flour beetle d) Ber beetle

37. **The commercial product 'Doom' has been developed from a**

a) Protozoan b) Fungus
c) Virus d) Bacterium

38. **The most successful field use of a baculovirus on global scale has been made for the control of**

a) Diamondback moth b) Cabbage caterpillar
c) Soybean caterpillar d) Pink bollworm

39. **The first commercial formulation of NPV was**

a) Gypchek b) Elcar
c) Virin- HS d) Spodopterin

40. **The share of microbial pesticides in the total world pesticide market is**
    a) 1-2% b) 5-6%
    c) 8-10% d) 12-15%

41. **The earliest documented case of insect resistance is against**
    a) Wheat stem sawfly b) Hessian fly
    c) Sorghum shoot fly d) Screw worm fly

42. **Major gene resistance is also called**
    a) Horizontal resistance b) Oligogenic resistance
    c) Vertical resistance d) Induced resistance

43. **DIMBOA is considered to be 31. resistance factor against European corn borer in**
    a) First generation b) Second generation
    c) Third generation d) Fourth generation

44. **The number of biotypes reported in brown plant hopper is**
    a) 2 b) 3
    c) 4 d) 5

45. **Which was the first country to commercialize transgenic crops?**
    a) USA b) China
    c) Canada d) Argentina

46. **The antifeedant properties of neem were first reported in India against**
    a) Tobacco caterpillar b) Desert locust
    c) American bollworm d) Brown plant hopper

47. **The irreversible inversion of acetyl cholinesterase is caused by**
    a) Malathion b) DDT
    c) HCH d) Carbaryl

48. **The first report of insecticide resistance in India was that of**
    a) Diamondback moth b) American bollworm
    c) Singhara beetle d) Mustard aphid

49. **The maximum amount of pesticides in India is used on.**
   a) Rice b) Vegetables
   c) Plantation crops d) Cotton

50. **The highest consumer of pesticides per unit area is**
   a) USA b) Taiwan
   c) Japan d) Korea

51. **Atropine sulphate is an**
   a) Antibiotic b) Antifeedant
   c) Antidote d) Antipheromone

52. **Methyl eugenol is an attract for**
   a) Oriental fruitfly b) Melon fruit fly
   c) Mediterranean fruitfly d) Paddy gall fly

53. **The level at which control measures should be initiated against a pest is called**
   a) Economic injury level
   b) Economic threshold level
   c) General equilibrium level
   d) None of these

54. **White backed plant hopper belongs to the family**
   a) Delphacidae b) Cicadellidae
   c) Coccidae d) Coreidae

55. **Rice leaf- folder belongs to the family**
   a) Noctuidae b) Pyralidae
   c) Arctiidae d) Pterophoridae

56. **Cotton whitefly is a vector of**
   a) Tungro virus b) Grassy stunt virus
   c) Leaf curl virus d) Leaf vein mosaic virus

57. **Diamondback moth belongs to the family**
   a) Pyralidae b) Yponomeutidae
   c) Pieridae d) Noctuidae

58. **Red pumpkin beetle belongs to the family**
   a) Chrysomelidae b) Curculionidae
   c) Dermestidae d) Bostrychidae

59. **Dengue fever is transmitted by**
   a) Anopheles spp. b) Culex spp.
   c) Mansonia spp. d) Aedes spp

60. **Amsacta moorei (Butler) is known as**
   a) Red hairy caterpillar b) Bihar hairy caterpillar
   c) Gram cutworm d) Greasy cutworm

61. **Epilachna vigintioclopunctata (Fabricius) is a pest of**
   a) Cucurbits b) Tomato
   c) Brinjal d) Lady's finger

62. **The Indian honey bee is technically named as**
   a) Apis dorsata Fabricius b) Apis florea Fabricius
   c) Apis cerana Fabricius d) Apis mellifera Linnaeus

63. **Who was awarded Nobel Prize for discovering the language of the honeybees?**
   a) Eva Crane b) Karl von Frisch
   c) S.P. Sakagami d) P. Ruttner

64. **Nosema disease of the honeybee is caused by a**
   a) Fungus b) Protozoan
   c) Virus d) Bacterium

65. **The fully ripened honey contains the highest amount of**
   a) Glucose b) Fructose
   c) Sucrose d) Water

66. **Royal gelly is the secretion of**
   a) Labial glands
   b) Neurosecretary cells

c) Hypopharyngeal glands
d) Mandibular glands

67. **The killing of pupae of silkworm in the cocoons is called**
a) Mounting b) Reeling
c) Curing d) Stifling

68. **Nosema bombycis causes a disease in silkworms known as**
a) Muscardine b) Pebrine
c) Flacherie d) Grasserie

69. **The rank of India in silk production in the world is**
a) First b) Second
c) Third d) Fourth

70. **The share of Rangeeni strain of lac insect to lac production in India is**
a) 10-20% b) 30-40%
c) 50-60% d) 80-90%

71. **The rank of India in lac production in the world is**
a) First b) Second
c) Third d) Fourth

72. **The Indian Lac Research Institute is situated at**
a) Hansi b) Ranchi
c) Pune d) Bangalore

73. **The first recipient of the World Food Prize was**
a) N.E. Borlaug b) B.P. Pal
c) R.F. Chandler d) M.S. Swaminathan

74. **The word 'Green revolution' was coined by**
a) William Gaud b) N.E. Borlaug
c) C. Subramaniam d) Robert McNamara

75. **The head quater of the World Food Prize Foundation is at**
   a) Rome (Italy) b) Des Moines (USA)
   c) Washington (USA) d) Manila (Philippine

Answers

1 d	2 c	3 b	4 b	5 d	6 d	7 a	8 c	9 b	10 a
11 c	12 b	13 c	14 b	15 d	16 c	17 b	18 a	19 d	20 a
21 b	22 c	23 b	24 a	25 b	26 c	27 b	28 d	29 c	30 b
31 c	32 a	33 c	34 d	35 a	36 b	37 d	38 c	39 b	40 a
41 b	42 c	43 a	44 d	45 b	46 b	47 a	48 c	49 d	50 b
51 c	52 a	53 b	54 a	55 b	56 c	57 b	58 a	59 c	60 a
61 c	62 b	63 b	64 b	65 c	66 d	67 b	68 b	69 b	70 d
71 b	72 a	73 d	74 a	75 b					

Chapter 10

Environmental Science

1. **The study of environment in relation to living organism is known as**
   a) Ecology b) Ecosystem
   c) Environment d) All of the above

2. **Human occupies central position in**
   a) Ecosystem b) Biosphere
   c) Stratosphere d) All of the above

3. **Which state has the largest proportion of net irrigated area under canal irrigation?**
   a) Punjab b) Uttar Pradesh
   c) Maharashtra d) Jammu and Kashmir

4. **Ecosystem is the structure and function of**
   a) Natural b) Artificial
   c) Environment d) Nature

5. **The non-degradable type of pollutant is**
   a) DET b) DEDP
   c) DDT d) Nitrogen

6. **Of the total water on the Earth, fresh water reserves constituent approximately–**
   a) 70% b) 2.7%
   c) 10% d) 8.9%

7. **Hydrosphere occupies 70% of**
   a) Global water b) Global land
   c) Forest d) All of the above

8. **Amongst the following Indian States which one has the minimum total forest cover**
   a) Sikkim b) Goa
   c) Haryana d) Kerala

9. **The linear pattern of eating and being eaten observed in nature is known as**
   a) Animal chain b) Food chain
   c) Human chain d) All of the above

10. **Which country has the maximum number of tube-wells in the world?**
    a) America b) Australia
    c) China d) India

11. **Energy flow and nutrient and function of**
    a) Ecosystem b) Environment
    c) Ecology d) All of the above

12. **Nuclear family refers to**
    a) Any family born after 1950
    b) Family includes parents and their children
    c) Entire family including children, their parents and grandparents
    d) Only husband and wife

13. **In troposphere temperature goes down up to**
    a) -50 C b) -65 C
    c) -60 C d) -55 C

14. **Which of the following is not an air pollutant?**
    a) Smoke b) Carbon Dioxide
    c) Nitrogen Gas d) Sulphur Dioxide

15. **O_3 gas protect life from harmful**
    a) Ultra violet radiation b) Infrared
    c) Visible d) All of the above

16. **Which part of plant evaporates water?**

a) Stomata b) Fruit
c) Branch d) Root

17. **Atmosphere layer lies 15 to 45 km above earth surface is**

a) Troposphere b) Biosphere
c) mesosphere d) Stratosphere

18. **Acid rain occurs due to oxides of**

a) Sulphur b) Nitrogen
c) Hydrogen d) Both A & B

19. **In elbow and knee joints, the bones can move in -**

a) All directions b) Four directions
c) One directions d) Two directions

20. **World forest day Is celebrated**

a) 21st March b) 22nd April
c) 5th July d) 16th October

21. **Which state has the largest area under tank irrigation?**

a) Tamil Nadu b) Odisha
c) West Bengal d) Karnataka

22. **Important GHG(ch4) is released from**

a) Soybean field b) Leguminous field
c) Paddy field d) Sorghum field

23. **The active factor for soil formation is**

a) Climate b) Time
c) Air d) Environment

24. **Pollination by wind is called-**

a) Autogamy b) Entomophily
c) Anemophily d) Ornithophily

25. **The energy required to evaporate one gram of water is**
    a) 540 Cal energy b) 680 Cal energy
    c) 580 Cal energy d) 768 Cal energy

26. **Time taken by sunlight to reach the earth surface**
    a) 6 min b) 8 min
    c) 7 min d) 12 min

27. **Gandhi Sagar Dam is a part of which one of the following?**
    a) Chambal Project
    b) Kosi Project
    c) Damodar Valley Project
    d) Bhakra Nangal Project

28. **The auditory effect of noise is**
    a) Deafness b) Dumb
    c) Blind d) All of the above

29. **The type of forest found maximum in India**
    a) Tropical dry deciduous b) Subtropical
    c) Climatic d) None of these

30. **Radioactivity is measured by–**
    a) Geiger-Muller counter b) Polarimeter
    c) Calorimeter d) Colorimeter

31. **The plant grow best at lower sunlight are known as**
    a) Phytophytes b) Microtherm
    c) Sciophytes d) None of these

32. **Condensation of moisture cause formation of**
    a) Smoke b) Smog
    c) Dew d) All of the above

33. **Which energy is converted into electrical energy by a solar cell?**
    a) Chemical Energy b) Nuclear Energy
    c) Solar Energy d) Magnetic Energy

34. **Plant require moderate temperature throughout year for growth is**

a) Mesotherm b) Micro therm
c) Mega therm d) None of these

35. **The cause of cancer is**

a) Edaphic b) Gasoline plant
c) Arsenic d) Carcinogen

36. **The sun total of all conditions that affects life and its development is known as**

a) Ecosystem b) Stratosphere
c) Environment d) All of the above

37. **Carrier of Dengue disease is**

a) Aedes Mosquito b) Culex Mosquito
c) Housefly d) Anopheles Mosquito

38. **Smog is formed due to reaction between fog and**

a) Smoke b) Dew
c) Hydrogen d) Nitrogen

39. **The main cause of global warming are**

a) CO b) CO_2
c) CH_4 d) Both B & C

40. **Wildlife Week is celebrated on**

a) 1st October to 7th October
b) 15th October to 2 1st October
c) 1st June to 7th June
d) 15th June to 21st June

41. **The instrument used to measure relative humidity**

a) Ozonometer b) Psychrometer
c) Anemometer d) Hygrometer

42. **In which state of matter, the distance between the molecules is minimum?**
    a) Solid b) Liquid
    c) Gas d) Plasma

43. **The gas leaked in Bhopal tragedy was**
    a) Methyl isocyanide b) Methyl cyanide
    c) Methyl nitro cyanide d) None of these

44. **Albedo of the earth is**
    a) 30% b) 40%
    c) 50% d) 60%

45. **Instrument is used for measuring the concentration of ozone in air**
    a) Psychrometer b) Anemometer
    c) Hygrometer d) Ozonometer

46. **Plants require high temperature throughout year for growth is classified as**
    a) Mesotherm b) Micro therm
    c) Megatherm d) Hekistotherm

47. **Maximum amount of gas found in air is**
    a) Oxygen b) Carbon dioxide
    c) Hydrogen d) Nitrogen

48. **Forest conservation act was passed in**
    a) 1988 b) 1980
    c) 1972 d) 1950

49. **Headache is related to**
    a) CO pollution b) NO_2 pollution
    c) CO_2 pollution d) SO_2 pollution

50. **The lowest temperature of air is recorded at–**
    a) Midnight b) Just before sunrise
    c) Just after sunset d) 2 a.m.

51. **Greenhouse gases absorbs**
 a) Long wave radiation
 b) Short wave radiation
 c) Medium wave radiation
 d) All of the above

52. **Atmospheric concentration of CO_2 is**
 a) 0.039% b) 0.025%
 c) 0.036% d) 0.30%

53. **Plants get their nitrogen from–**
 a) Rain b) The soil
 c) The air d) The bedrock

54. **The world environment day is celebrated on**
 a) 5th June b) 21st March
 c) 22nd April d) 5th July

55. **Which one among the following covers the highest percentage of forest area in the world?**
 a) Temperate Coniferous forests
 b) Temperate Deciduous forest
 c) Tropical Monsoon forests
 d) Tropical Rain forests

56. **Atmospheric concentration of N_2 is**
 a) 76% b) 21%
 c) 78% d) 68%

57. **Atmospheric concentration of O_2 is**
 a) 23% b) 21%
 c) 29% d) 33%

58. **Which of the following is not a primary pollutant?**
 a) SO_2 b) Volcanic ash
 c) O_3 d) CO_2

59. **The water disinfectant is**
    a) Chlorine b) Iron
    c) Magnesium d) Chloride

60. **Which state irrigates largest percentage of land?**
    a) Haryana b) Uttar Pradesh
    c) Maharashtra d) Punjab

61. **Land that can be cultivated to grow crop is known as**
    a) Arable land b) Cultivated land
    c) Command area d) All of the above

62. **Bhopal disaster was occurred in**
    a) 1985 b) 1984
    c) 1992 d) 1990

63. **Which of the following is used as a moderator in nuclear reactor?**
    a) Thorium b) Graphite
    c) Radium d) Ordinary water

64. **Blue baby syndrome is caused by higher concentration of**
    a) NO_3 b) SO_2
    c) CO_2 d) Mg

65. **What unit of solar energy does the earth intercept?**
    a) 2,200 millionth b) 2,000 millionth
    c) 2,100 millionth d) 2,300 millionth

66. **Bronchial spasm is related to**
    a) NO_2 pollution b) CO_2 pollution
    c) SO_2 pollution d) CO pollution

67. **The most harmful environmental pollution from nuclear reactor is**
    a) Radioactivity b) Particulate formation
    c) Thermal pollution d) Noise pollution

68. **Maximum concentration of ozone is found in**
   a) Stratosphere b) Troposphere
   c) Mesosphere d) Ionosphere

69. **Bronchial fibrosis is related to**
   a) NO_2 pollution b) CO_2pollution
   c) SO_2 pollution d) CO pollution

70. **The biochemical laboratory of human body is**
   a) Stomach b) Liver
   c) Intestine d) Kidney

71. **Density of water is maximum at**
   a) 5 C b) 6 C
   c) 10 C d) 4 C

72. **The colour less odourless and non-corrosive air pollutant**
   a) SO_2 b) NO_2
   c) CO_2 d) None of these

73. **Which state of India is known as 'Tiger State'?**
   a) Gujarat b) West Bengal
   c) Madhya Pradesh d) Assam

74. **Earth day is celebrated on**
   a) 22nd April b) 21st March
   c) 16th October d) 5th June

75. **Fungi are plant that lack**
   a) Chlorophyll b) Chloroplast
   c) Photochemical d) lignin

76. **Which of the following has maximum speed?**
   a) Air b) Water Current
   c) Sound d) Light

77. **Wild life protection act was passed in**
    a) 1975 b) 1972
    c) 1980 d) 1995

78. **In polluted lake index of pollution is**
    a) Daphnia b) Edaphic
    c) Predation d) None of these

79. **Instrument used for wind speed**
    a) Psychrometer b) Anemometer
    c) Hygrometer d) Ozonometer

80. **Tobacco addiction is caused due to**
    a) Cocaine b) Caffeine
    c) Nicotine d) Histamine

81. **Global warming is mainly due to**
    a) Greenhouse gases b) Co
    c) N_2 d) H_2

82. **Comparative content of material is less in municipal solid waste is**
    a) Non-recyclable b) Recyclable
    c) Partially recyclable d) None of these

83. **Wavelength of infrared radiations is**
    a) Zero b) Finite
    c) Shorter d) longer

84. **The predominate forests in the world is**
    a) Tropical forest b) Subtropical forest
    c) Rain forest d) None of these

85. **Which one of the following is not a Biosphere Reserve?**
    a) Agasthyamalai b) Nallambalai
    c) Nilgiri d) Panchmarhi

86. **Permissible limit of iron in drinking water is**

a) 2 ppm b) 1 ppm
c) 3 ppm d) 5 ppm

87. **The microbes that require o2 for their existence is**

a) Anaerobic b) Aerobic
c) Both A & B d) NoOne of these

88. **Largest salt water lake in India**

a) chilka b) Lonar
c) Dal d) Sukhna

89. **Nitrous oxide (commonly called laughing gas) has been a matter of concern to environmentalist recently because-**

a) It is thought to cause cancer at low concentration
b) It produce photochemical smog
c) It is a greenhouse gas
d) None of the above

90. **Biodiversity act was proposed in**

a) 2000 b) 1991
c) 1980 d) 1972

91. **In grassland ecosystem producers are of**

a) Large size b) Small size
c) Medium size d) None of these

92. **The soil of India's eastern and western coast is–**

a) Alluvial b) Black cotton
c) Red rocky d) Laterite

93. **Zooplanktons are primary consumers in**

a) Aquatic system b) Aeroponic
c) Hydroponic d) All of the above

94. **Acid rain pH is**

a) 5.8 b) 6.8
c) 6.5 d) 5.6

95. **When an ant bites a person, which irritating chemical it (ant) injects into his (person) body?**

a) Acetic acid b) Citric acid
c) Tartaric acid d) Formic acid

96. **The national park where 'lion' are being conserved is**

a) Kanhakishli b) Gir
c) Cobett d) All of the above

97. **The rate of solubility of oxygen in water at 30 is**

a) 7.6 mg/l b) 8.5 mg/l
c) 5.6 mg/l d) 8.2 mg/l

98. **How much of the net irrigated well irrigation-**

a) 20% b) 60%
c) 45% d) 30%

99. **The project tiger was launched in**

a) 1975 b) 1972
c) 1980 d) 1985

100. **The height of troposphere from equator is**

a) 15 km b) 18 km
c) 20 km d) 32 km

101. **Number of atoms in ozone molecules are**

a) 2 b) 3
c) 4 d) 1

102. **For Analgesic use**

a) Morphine b) Chlorine
c) Minamata d) None of these

103. **Higher energy level and shorter wavelengths are features of**

a) Beta radiation b) Alpha radiation
c) Ultraviolet radiation d) Infrared radiation

104. **The unit of measure of concentration of ozone is**

a) Dobson b) Ozonometer
c) Anemometer d) Hygroscopy

105. **Pyramid of energy is always**

a) Down right b) Straight
c) Triangle d) Upright

106. **Chemical released by chlorofluorocarbons that causes depletion of ozone layer in atmosphere is**

a) Nitrogen b) Sulphuric acid
c) Chlorine d) Sodium chloride

107. **Minamata disease is caused due to mercury metal in**

a) Drinking water b) Fresh water
c) Sewage water d) Sea water

108. **The maximum permissible limit of nitrate in drinking water**

a) 60 mg/l b) 80 mg/l
c) 50 mg/l d) 45 mg/l

109. **When chlorofluorocarbons are released in atmosphere, chlorine chemical destroys the**

a) Beta molecules b) Helium molecules
c) Argon molecules d) Ozone molecules

110. **The solar radiation used in photosynthesis is**

a) 3% b) 4%
c) 2% d) 5%

111. **The instrument is used for sound intensity measurement**

a) Sonometer b) Anemometer
c) Sound meter d) None of these

112. **Precipitation occur in form of minute drops float in air is known as**

a) Rain b) Dew
c) Smoke d) Drizzle

113. **Lower energy level and longer wavelengths are features of**
   a) Ultraviolet radiation b) Infrared radiation
   c) Beta radiation d) Alpha radiation

114. **The forest, rich in biodiversity is**
   a) Sub-tropical b) Evergreen
   c) Tropical forest d) All of the above

115. **The liquid gold is known as**
   a) Petroleum b) Liquid nitrogen
   c) Water d) All of the above

116. **Most serious group of chemicals emitted in British Columbia that causes depletion of ozone layer are**
   a) Noble gases b) Halons
   c) Halogens d) Helium

117. **The factor responsible for stomata opening and closing**
   a) Light b) Climate
   c) CO_2 d) Water

118. **The indicator of organism pollution in water is**
   a) DOC b) BOD
   c) COD d) Both B & C

119. **Height of ozone above surface of Earth is**
   a) 30 to 50 km b) 10 to 20 km
   c) 15 to 30 km d) 50 to 70 km

120. **The plant that produce food themselves through photosynthesis are**
   a) Microtherm b) Amino acid
   c) Photocabon d) Phototrophs

121. **The most poisonous water pollutant is**
   a) Ca b) Mg
   c) N d) Ar

122. **The first biosphere reserve in india was**

a) Nilgiri
b) Sunderbans
c) Gulf of mannar
d) Nokrek

123. **Burning of fossil fuels**

a) Decrease greenhouse gases
b) Increases greenhouse gases
c) Increased level of oxygen
d) Increased level of ethane

124. **The temperature in mesosphere reach as low as**

a) -90 C
b) -80 C
c) -50 C
d) -60 C

125. **Name of protocol signed to reduce production of chlorofluorocarbons is**

a) UVB protocol
b) Montreal Protocol
c) UVA protocol
d) UVR protocol

126. **The main constituent of biogas is**

a) CH_4
b) NO_2
c) SO_2
d) CO

127. **The distinct zones of atmosphere is**

a) 9
b) 6
c) 5
d) 7

128. **Annual ozone hole is located on continent**

a) Antarctica
b) South America
c) North America
d) Africa

129. **The main driving force of hydrological cycle is**

a) Solar radiation
b) Climate
c) Environment
d) Atmospher

130. **Vermicomposting is done by**

a) Fungus
b) Bacteria
c) Worms
d) Animal

131. **Rate of photosynthesis is higher in**
   a) Interment light b) Long wave length
   c) Short wave length d) Red light

132. **The temperature generally decrease with an ……… in altitude**
   a) Increase b) Decrease
   c) Above d) None of these

133. **World population day is celebrated on**
   a) 11th July b) 5th June
   c) 21st March d) 22nd April

134. **Cutting of trees in a forest is called :**
   a) Deforestation b) Afforestation
   c) Reforestation d) None of these

135. **Photo chemical smog occurs mainly due to reaction between**
   a) Sunrays b) NO2
   c) CO2 d) Both A & B

136. **World nature day is celebrated on**
   a) 3rd October b) 2nd October
   c) 5th July d) 7th March

137. **The sound of which intensity is regarded as the safe noise level for a city?**
   a) 80 dB b) 60 dB
   c) 120 dB d) 45 dB

138. **CFC s gas was banned in**
   a) Montreal protocol b) Kyoto protocol
   c) Lock protocol d) All of the above

139. **In the type of interaction where one organism gets benefited and another harmed is known as**
   a) Penicillium b) Phototrophs
   c) Parasitism d) None of these

140. **World health day is celebrated on**
   a) 7th April
   b) 2^{nd}October
   c) 5th July
   d) 7th March

141. **Which of the following disease is not caused by water pollution?**
   a) Malaria
   b) Jaundice
   c) Typhoid
   d) Dysentery

142. **The most important cause for the loss of biodiversity is**
   a) Climate change
   b) Atmosphere change
   c) Wind speed change
   d) All of the above

143. **Lead in water can cause**
   a) Kidney damage
   b) lever
   c) lungs
   d) All of the above

144. **Plants are killed at low temperature because :**
   a) Desiccation takes place owing to the withdrawal of water from vacuolated protoplasm
   b) Precipitation of cell proteins
   c) Cells rupture due to the mechanical pressure of ice
   d) All the above three are correct

145. **The solar radiation that reaches the earth's surface is**
   a) 45-58%
   b) 40-50%
   c) 50-60%
   d) 60-75%

146. **The commodities considered valuable in their relatively unmodified form is**
   a) Natural resources
   b) Ecosystem
   c) Artificial resources
   d) None of these

147. **Soil erosion can be prevented by :**
   a) Increasing bird population
   b) Afforestation
   c) Removal of vegetation
   d) Over grazing

148. **Component of soil imparts to most of the life sustaining attributes is**

a) Minerals b) Sulphur

c) Phosphors d) Organic matter

149. **The target organ of cadmium toxicity is**

a) Lever b) Kidney

c) Lungs d) Eyes

150. **Spraying of DDT on crops causes pollution of**

a) Soil and Water b) Air and Soil

c) Crops and Air d) Air and Water

151. **The highly exploited natural resource is**

a) Water b) Soil

c) Air d) All of the above

152. **The first national forest policy was formulated in**

a) 1958 b) 1965

c) 1952 d) 1980

153. **CFC was discovered first as**

a) CFC-11 b) CFC-12

c) CO_2 d) NO_2

154. **Hardness of water is due to presence of**

a) $Caco_4$ b) $Caco_3$

c) $Mgco_3$ d) Both B & C

155. **Euphorbia lathyris is a**

a) Gasoline plant b) Khejri tree

c) Mesotherm plant d) None of these

156. **Founder of Kenyan greenbelt movement is**

a) Wangari maathai b) Bergman

c) Charles Elton d) None of these

157. **Corbett national park located in the state Uttarakhand is famous for**
   a) Tiger b) Elephant
   c) Golden jackal d) lion

158. **Atmospheric humidity is measured by**
   a) Hygrometer b) Hydrometer
   c) Hygroscopy d) None of these

159. **Proteins are made up of**
   a) Carbohydrate b) Fat
   c) Amino acids d) Vitamins

160. **Plants that require low temperature throughout the year for growth is classified as**
   a) Microtherm b) Mesotherm
   c) Hekistotherm d) None of these

161. **Noise is the**
   a) Unwanted sound b) Sound
   c) Sound wave d) All of the above

Answers

1	a	2	b	3	d	4	a	5	c	6	b	7	a	8	c	9	b	10	d
11	a	12	b	13	c	14	c	15	a	16	a	17	d	18	d	19	c	20	a
21	a	22	c	23	a	24	c	25	c	26	b	27	c	28	a	29	a	30	a
31	c	32	c	33	c	34	a	35	d	36	c	37	a	38	a	39	d	40	a
41	b	42	a	43	a	44	a	45	d	46	c	47	d	48	b	49	a	50	b
51	a	52	c	53	b	54	a	55	a	56	c	57	b	58	c	59	a	60	d
61	a	62	b	63	b	64	a	65	b	66	c	67	c	68	a	69	a	70	b
71	d	72	a	73	c	74	a	75	a	76	d	77	b	78	a	79	b	80	c
81	a	82	b	83	d	84	c	85	b	86	b	87	b	88	b	89	c	90	a
91	a	92	d	93	a	94	d	95	d	96	b	97	a	98	d	99	b	100	a
101	b	102	a	103	c	104	a	105	d	106	c	107	a	108	c	109	d	110	c
111	a	112	a	113	d	114	b	115	c	116	a	117	b	118	a	119	d	120	c
121	d	122	d	123	a	124	b	125	b	126	b	127	a	128	c	129	a	130	a

131 b	132 a	133 a	134 a	135 b	136 d	137 a	138 d	139 a	140 c
141 a	142 a	143 a	144 a	145 d	146 b	147 a	148 b	149 d	150 b
151 a	152 c	153 c	154 b	155 d	156 a	157 a	158 a	159 a	160 c
161 a	162 a								

❑❑❑

Chapter 11

Genetics and Plant Breeding

1. **Self-pollinated species are also known as___________ species or inbreeders.**
   a) Allogamous b) Cleistogamy
   c) Chasmogamy d) Autogamous

2. **__________ has produced vsuccessful hybrids through artificial crosses.**
   a) Camerarius b) Fairchild
   c) Koelreuter d) Jhonson

3. **__________ is a progeny of single self-pollinated homozygous plant.**
   a) Female parent b) Pureline
   c) Inbred d) Hybrid

4. **Plant breeding is a science and art and technology which deal with ___________improvement of crop plants in relation to their economic use for mankind.**
   a) Yield b) Production
   c) Genetic d) Good

5. **_______________ method of breeding is not appropriate for cross pollinated crops.**
   a) Mass Selection b) Back cross method
   c) Heterosis breeding d) Pure line selection

6. **_____________ is an ancient method of crop improvement.**
   a) Selection b) Introduction
   c) Hybridization d) Mutagenesis

7. **___________is the oldest selection methods for crop improvement**

a) Mass selection b) Pure line selection

c) Pedigree selection d) Recurrent selection

8. **____________ is a not a general breeding method**

a) Introduction b) Selection

c) Distant hybridization d) Hybridization

9. **____________ Pollinated crops have homozygous balance and are tolerent to inbreeding.**

a) Self b) Cross

c) Often cross d) Non of all

10. **Variety developed by mass selection method is mixture of homozygous and heterozygous plants and population is heterogeneous in _________________pollinated crop**

a) Self b) Cross

c) Often cross d) None of these

11. **_______breeding refers to selection procedure in which the segregating population of self pollinated species is grown without selections.**

a) Hybridization b) Pedigree

c) Bulk d) Back cross

12. **The self-progeny of F1 and the subsequent generations are termed as _______________generations.**

a) F_2 b) Segregating

c) Advance d) Selfed

13. **_________ breeding method is generally used to improve specific character of a well adapted variety**

a) Backcross b) Pedigree

c) Bulk d) Hybridization

14. **Cross between two genetically different homozygote plants is __________**

a) Progeny b) Variety
c) Back cross d) Hybrid or F1

15. **__________method does not provide opportunity to practice selection for superior plant till F5 generations.**

a) Single seed descent method
b) Back cross method
c) Heterosis breeding
d) Pure line selection

16. **____________method provides information about the mode of inheritance of various qualitative characters which is not possible by other breeding methods.**

a) Back cross b) Pedigree
c) Hybridization d) Mutagenesis

17. **__________ is not a useful method for handling segregating populations**

a) Pedigree breeding b) Bulk breeding
c) Back cross breeding d) Hybridization

18. **___________ Parent is used only once in back cross breeding method**

a) Recurrent b) Female
c) Male d) Donor

19. **_________ method is most useful for study competitive ability of genotypes in population.**

a) Mass Selection b) Back Cross breeding
c) Hybridization d) Non of all

20. **In pedigree breeding after F_8, homozygous plants are known as __________**

a) Varieties b) Strains
c) Genotypes d) Segregants

21. **Cross pollinated species are also known as__________ species or inbreeders.**

a) Allogamous b) Cleistogamy
c) Chasmogamy d) Autogamous

22. __________ **is a progeny of two different single self-pollinated homozygous plants.**

a) Female parent b) Pureline
c) Inbred d) Hybrid

23. ______________ **method of breeding is appropriate for improvement of good variety.**

a) Mass Selection b) Back cross method
c) Heterosis breeding d) Pure line selection

24. __________**is the newer methods for crop improvement**

a) Selection b) Introduction
c) Hybridization d) Mutagenesis

25. __________ **is a general breeding method**

a) Hybridization b) Mutation
c) Distant hybridization d) Introduction

26. __________ **pollinated crops have heterozygous balance and are not tolerant to inbreeding.**

a) Self b) Cross
c) Often cross d) Non of all

27. **Mendel theorized that genetic traits are "segregated" during gamete formation and the offspring get only one gene for a trait from each parent. Why is this important to sexually reproducing organisms?**

a) They would get too many dominant traits if it did not occur
b) They would get too many recessive traits if it did not occur
c) It allows for offspring to be different from their parents
d) It allows the best traits to be selected out for the offspring

28. **Mendel theorized that genetic traits are "independently assorted" and one trait doesnot depend on another for transmission to offspring. What do we know today that makes this theory invalid?**
    a) Genes on the same chromosome are usually transmitted together.
    b) Chromosomes stick together and are often transmitted with each other.
    c) The same gene may be responsible for several different traits.
    d) Transmission of genes is random and depends on laws of probability

29. **Where are genes for sex-linked traits located?**
    a) They can be on any chromosome
    b) On the "Y" chromosome
    c) On the "X" chromosome
    d) On chromosome pair # 21

30. **White flower crossed with a red flower produces pink offspring. This is an example of incomplete dominance. If two of the pink-flowered plants are crossed, what ratio will the genotypes of the offspring have?**
    a) 1 RR: 3 Rr
    b) 1 RR: 2 RR′: 1 R′R′
    c) 2 Rw: 2 Ww
    d) 1 Rr: 2 rr: 1 ww

31. **Why blood types are considered an example of codominance?**
    a) there are four types of blood, not two
    b) blood type O can be donated to anyone, AB cannot
    c) blood types A and B will both be expressed when present
    d) there are three alleles for blood type and four types of blood

32. **How many pairs of homologous chromosomes do humans have?**
    a) 4
    b) 23
    c) 44
    d) 46

33. **If two different alleles are present, which allele is expressed?**
    a) recessive
    b) dominant
    c) phenotypic
    d) pure

34. **Tall pea plant (D) and a tall pea plant (D) have what in common?**
    a) phenotype b) genotype
    c) alleles d) seed color

35. **In crossing homozygous dominant and homozygous recessive pea plants, Mendelnoted that some genes were not seen in the F1 generation and were seen in only 25%of the F2 generation. What did he call these genes?**
    a) dominant b) recessive
    c) lethal d) incompletely dominant

36. **Continuous inbreeding (Selfing) leads**
    a) Homozygosity b) Hetrozygosity
    c) Both d) None

37. **Composite varieties are developed in crops**
    a) Selfpollinated b) Cross pollinated
    c) Clone d) All of these

38. **Broader genetic base is found in**
    a) Mass selaction b) Pureline selection
    c) Clonal selection d) All of these

39. **Effective selection can be made in**
    a) Pureline b) Hybrid
    c) Clone d) Multilines

40. **Heterosis results due to complementation between**
    a) Different genes b) Divergent alleles
    c) Multiple alleles d) Iso - alleles

41. **Pollinaton and fertilization occers before opening of flower is termed as**
    a) Chasmogamy b) Cliestogamy
    c) Homogamy d) Hetrogamy

42. **Pollens are not produced or production of sterile pollens by a plat is termed as**
   a) Self incompatibility b) Male sterility
   c) Inability d) All of these

43. **Crops are classified as Cross and self pollinated on the basis of mode of**
   a) Reproduction b) Pollination
   c) Growth d) All of these

44. **Mixture of iso-line is termed as**
   a) Synthetic b) Composite
   c) Pureline d) Multiline

45. **Vegetative embryos develops without fertilization**
   a) Seed b) Apomixis
   c) Monoacy d) hermaphrodity

46. **Female parent of world first CGMS based Pigeon pea hybrid**
   a) MS(P)DT b) GT288A
   c) GTR11 d) GTH-1

47. **CGMS system developed in the crop**
   a) Greengram b) Pigeon pea
   c) Garlic d) termaric

48. **Multiline breeding is exploited widly in the crop**
   a) Wheat b) Cotton
   c) Ginger d) Pigeon pea

49. **Hybridiztion is common in crop**
   a) Wheat b) Cotton
   c) Ginger d) Sorghum

50. **Clonal selection mostly used in the crop**
   a) Wheat b) Cotton
   c) Ginger d) Sorghum

51. **Term heterosis given by**

a) Kelruter b) Mendel

c) Shull d) Borloag

52. **Father of Genetics**

a) Kelruter b) Mendel

c) Shull d) Borloag

53. **Father of green revolution of india**

a) Swaminathan b) Bourloug

c) Raddy d) Shrivastava

54. **Genetic purity is maintained by**

a) Roughing b) Authatic seed source

c) Isolation d) All of these

55. **Progeny of breeder seed is**

a) Nucleous seed b) Foundation seed

c) Breeder Seed d) Certified seed

56. **Seed act formulated and anacted during**

a) 1966 and 1969 b) 1967 and 1970

c) 1967 and 1968 d) 180 and 1982

57. **The standerd method of seed moisture estimation**

a) Oven dry b) Toulene

c) Moisture meter d) P_2O_5

58. **Seed certification in Gujarat is done by**

a) GSSCA b) GSSC

c) GCCS d) GSRC

59. **Color of Breeder seed certificate tag is**

a) Golden red b) Golden Green

c) Golden Yellow d) Ligh Red

60. **Hand pollination during commercial seed production is practiced in**

a) Sorghum b) Pigeonpea
c) Cotton d) Cator

61. **Protection period for field crop varieties in plant breeder's right is**

a) 10 Years b) 15 Years
c) 7 Years d) 3 Years

62. **UPOV Head quarter is at**

a) Rome b) Geneva
c) New York d) Tokyo

63. **Isolation distance in Crtified seed production of Bajara is in meter**

a) 100 b) 150
c) 200 d) 50

64. **Source of breeder seed is**

a) Certified b) Nucleous
c) Foundation d) Registered

65. **Sceientis has made sugession to maintain genetic purity of variety**

a) Mendel b) Shull
c) Horne d) Borloag

66. **Certified seed tag having**

a) Yellow color b) Red color
c) Blue color d) White color

67. **Seed prosesing refers to**

a) Drying b) Packing
c) Grading d) All of these

68. **Seed testing refers to**

a) Purity test b) Seed moisture test
c) Germination test d) All of these

69. **Varietal purity is checked by**
    a) Grow out test b) Accelerated aging test
    c) Tz Test d) Germination test

70. **The main objective of field inspection is to examine**
    a) Disease incidence b) Isolation distance
    c) Offtypes d) All of these

71. **Authority advices the central and state govt. on all matters related to seed**
    a) Central seed committee b) Seed testing laboratory
    c) ICAT d) Agricultural University

72. **The scientific manipulation of living organisms, especially at the molecular genetic level, to produce useful products. Gene splicing and use of recombinant DNA (rDNA) are major techniques used.**
    a) Breeding b) Biotechnology
    c) Tissueculture d) Genetics

73. **Pulse polio vaccine is**
    a) Killed virus b) Live virus vaccin
    c) Virus protiens d) all

74. **A virus that in- fects bacteria. Altered forms are used as vectors for cloning DNA.**
    a) Host b) Virulent
    c) Viroid d) Bacteriophase

75. **A bacterium that kills insects; a major component of the microbial pesticide industry.**
    a) Agrobacterium b) Bacillus thuranginsis
    c) Biokiller d) Spirobacteria

76. **A library composed of complementary copies of cellular mRNAs**
    a) Central b) cDNA
    c) Reserve d) Protected

77. **Usually the location in the world where the oldest cultivation of a particular crop has been identified.**

a) Center of inovation b) Center of research

c) Center of Gene d) Center of origin

78. **Francis Crick's seminal concept that in nature genetic information generally flows from DNA to RNA to protein.**

a) Gene expiration b) Central Dogma

c) Coding d) DNA-RNA-Protienbredging

79. **The central portion of the chromosome to which the spindle fibers attach during mitotic and meiotic division.**

a) Nucleous b) Centromere

c) Center d) Telomere

80. **A) single DNA molecule, a tightly coiled strant of DNA, condensed into a compact structure in vivo by complexing with accessory histones or histone-like proteins.**

a) Genome b) Genepool

c) GenomicDNA d) Chromosome

81. **A) DNA sequence that codes for a specific polypeptide;**

a) Cistron b) Gene

c) allel d) Codon

82. **An exact genetic replica of a specific gene or an entire organism.**

a) Duplicate b) Clone

c) Copy organism d) Iso organism

83. **Members of the pairs adenine-thymine, adenine-uracil, and guaninecytosine that have the ability to hydrogen bond to one another.**

a) Nucleotide

b) Nucleoside

c) Complimentory nucleotide

d) Codon

84. **A) cell which contains genetically identical two copies of each chromosome**

a) Haploid b) Diploid
c) Triploid d) Doubble haploid

85. **Procedures for determining the nucleotide sequence of a DNA fragment.**

a) DNA finger printing b) DNA Sequencing
c) Replication d) DNA amplification

86. **The organisms in a plant population and the biotic and abiotic factors which impact on them.**

a) Ecosystem b) Ecology
c) Factor d) Biom

87. **A) method for transforrning DNA, especially useful for plant cells, in which high voltage pulses of electricity are used to open pores in cell membranes, through which foreign DNA can pass**

a) Electroporation b) Electrophorosis
c) Electrolysis d) Electrodigetion

88. **A) haploid sex cell, egg or sperm, that contains a single copy of each chromosome.**

a) Sex cell b) Gamate
c) Zygote d) Embryo

89. **A) locus on a chromosome that encodes a specific protein or several related proteins. It is considered the functional unit of heredity**

a) Factor b) Allele
c) Gene d) Gene sheet

90. **The process of producing a protein from its DNA- and mRNA-coding sequences.**

a) Protein production b) Ancoding
c) genetics d) Gene expresion

91. **The manipulation of an organism's genetic endowment by introducing or eliminating specific genes through modern molecular biology techniques.**

a) Biotechnology b) Tissue structurin

c) Genology d) Genetic engineering

92. **A) linear map of the relative positions of genes along a chromosome.**

a) Gene map b) Linkage map

c) Genetic code d) dandogram

93. **A) gene or group of genes used to "mark" or track the action gene.**

a) Gene to gene b) allele

c) Action controler d) **Genetic marker**

94. **A) category including closely related species. Interbreeding between organisms within the same category can occur.**

a) Genus b) Family

c) Kingdom d) Tribe

95. **An organism whose genotype is characterized by two identical alleles of a gene.**

a) Hetrozygote b) Homozygote

c) Hemizygote d) Zygote

96. **A) chart that traces the flow of genetic information from generation to generation.**

A) record B) linkage

C) lineage D) generations

97. **The frequency of coinheritance of a pair of genes and/or genetic markers, which provides a measure of their physical proximity to one another on a chromosome.**

A) Linkage B) Map

C) Gene map D) Pedigree

98. **Determining the physical location of a gene or genetic marker on a chromosome.**
   a) marking b) Siting
   c) Mapping d) Taging

99. **DNA amplification is done in machine.**
   a) Thermocycler b) Incubator
   c) Hybridization oven d) Electrophoresis system

100. **The reduction division process by which haploid gametes and spores are formed, consisting of a single duplication of the genetic material followed by two mitotic divisions.**
   a) Cell division b) Mitosis
   c) Cell division d) Mieosis

Answers

1 d	2 b	3 b	4 c	5 d	6 b	7 a	8 c	9 a	10 a
11 c	12 b	13 a	14 d	15 a	16 b	17 b	18 d	19 c	20 b
21 a	22 d	23 b	24 d	25 d	26 b	27 c	28 c	29 b	30 b
31 d	32 b	33 b	34 a	35 b	36 a	37 b	38 a	39 d	40 d
41 b	42 b	43 b	44 d	45 b	46 b	47 b	48 a	49 b	50 c
51 c	52 b	53 a	54 d	55 b	56 a	57 a	58 a	59 c	60 c
61 b	62 b	63 b	64 b	65 c	66 c	67 c	68 d	69 a	70 d
71 a	72 b	73 b	74 d	75 b	76 b	77 d	78 b	79 b	80 d
81 a	82 b	83 c	84 d	85 c	86 a	87 a	88 b	89 c	90 d
91 d	92 c	93 d	94 a	95 b	96 c	97 a	98 c	99 a	100 d

❑❑❑

Chapter 12

Horticulture

1. **De-cap medicine for diabities prepared from**
   a) Bougainvillea b) Rose
   c) Dahalia d) Mussenda

2. **Pusa Urmi and Pusa Shreyali are the varieties of**
   a) Pointed gourd b) Ash gourd
   c) Bitter gourd d) Snake gourd

3. **Fruit dormancy is noticed in**
   a) Apple b) Annona
   c) Wood apple d) Aonla

4. **Botanically, aonla fruit is a**
   a) Capsule b) Nut
   c) Pome d) Berry

5. **Origin of Aonla is**
   a) China b) India
   c) Tropical America d) South America

6. **Most suitable aonla variety for candy making is**
   a) NarendraAonla – 6 b) Banarasi
   c) Chakkaiya d) Francis

7. **Which of the following anola variety which highly susceptible to fruit necrosis disorder?**
   a) Francis b) Chakaiya
   c) BSR-1 d) Banarasi

8. **Anti-oxidant property of aonla is due to presence of**
   a) Tartaric acid b) Mallic acid
   c) Gallic acid d) Ascorbic acid

9. **The fruit crop which can be grown in wide range of sodic soil is**
   a) Papaya b) Aonla
   c) Guava d) Sapota

10. **Three tier fruit crop model system which suits for sodic soil is**
    a) Aonla + Ber + Phalsa b) Aonla +Ber+ Papaya
    c) Aonla +Ber+ Guava d) Aonla +Ber+ Sapota

11. **Fruit necrosis in aonla is due to– – – – – – – –.**
    a) 'B' deficiency b) 'Ca' deficiency
    c) 'K' deficiency d) 'Zn' deficiency

12. **– – % of fruits droped in aonla within three weeks of flowering.**
    a) 70% b) 20%
    c) 50% d) 10%

13. **Aonla is best suited to grow under which one of the following soil condition?**
    a) Acidic soil b) Slightly acidic to saline sodic soil
    c) Calcareous soil d) Alkali soil

14. **Which one of the following banana cultivar has got "Geographical Indication" status?**
    a) Nendran b) Hill banana
    c) Rasthali d) Robusta

15. **Which one of the following is a climacteric type of fruit?**
    a) Banana b) Citrus
    c) Litchi d) Grapes

16. **Somatic chromosome number of 'Grand Naine' banana is**
    a) 11 b) 22
    c) 33 d) 44

17. **---------- is the culinary variety of banana**

a) Poovan b) Rasthali

c) Monthan d) Robusta

18. **----------- is the fruit crop that occupies the highest area in India**

a) Mango b) Banana

c) Citrus d) Grapes

19. ***Musa balbisiana* is characterized by both**

a) Slight pseudostem blotches and downy peduncle

b) Heavily marked pseudostem blotches and downy peduncle

c) Slight pseudostem blotches and glabrous peduncle

d) Heavily marked pseudostem blotches and glabrous peduncle

20. **A tetraploid dessert banana with apple flavor and**

a) FHIA-1 b) FHIA-3 downy peduncle

c) FHIA-21 d) Red banana

21. **In which of the following banana cultivar male bud is absent ?**

a) French Plantain b) Horn Plantain

c) Williums d) Red Banana

22. **Polyploid breeding is common in ----------- fruit crop**

a) Banana b) Apple

c) Papaya d) Citrus

23. **The explant used for commercial tissue culture of banana is**

a) Leaf bits b) Stem bits

c) Shoot tips d) Petiole

24. **The genomic constitution of Cavendish group of banana is**

a) AAA b) AAB

c) ABB d) AA

25. **The operation in banana where 60 cm of the mother plant from the ground level after harvest is retained for nutrient supply to daughter suckers is**

a) Desuckering b) Denavelling

c) Mattocking d) Propping

26. **———————state is the leader in banana production in India**

a) Uttar Pradesh b) Maharastra

c) Andhra Pradesh d) Tamil Nadu

27. **The spacing recommended for HDP(3 suckers/clump) in banana by TNAU is**

a) 1.8m x 1.8m b) 2.0m x 2.0m

c) 3.6m x 1.8m d) 1.8m x 3.6m

28. **Which of the following is banana hybrid**

a) Co-1b b) PKM-1

c) Poovan d) IIHR -1

29. **Banana is otherwise called as**

a) King of the fruit b) Queen of the fruit

c) Apple of Paradise d) Prince of fruit

30. **Which of the following is triploid*accuminata* commercial variety of banana?**

a) Grand Naine b) Nendran

c) Poovan d) Virupaksh

31. **Banana is commercially propagated through**

a) Water suckers b) Sword suckers

c) Peepers d) Tissue culture of floral buds

32. **Removal of male bud after completion of female phase in banana is referred as**

a) Tipping b) Denavelling

c) Thinning d) Pinching

33. **The banana fruits are packed in 100 gauge polybags with following % holes to increase the shelf life**

a) 0.2% b) 2.0%

c) 5.0% d) 10.0%

34. **Ney poovan variety of banana belongs to ------------- genome group of banana.**

a) AA b) AAA

c) AB d) AAB

35. **Paring and Pralinage is followed in banana to control**

a) Aphids b) Stem borer

c) Nematodes d) Fusarium wilt

36. **Tissue culture propagation is gaining importance in ---------- crop.**

a) Citrus b) Grape

c) Banana d) Papaya

37. **--------- is the one of the popular parthenocarpic fruits.**

a) Banana b) Mango

c) Citrus d) Grapefruit

38. **Storage temperature for banana is --------.**

a) 5-10°C b) 10-15°

c) 13°C d) 20-21°C

39. **The major problem in banana breeding is -----------**

a) Long duration

b) Seedlessness

c) Complex flower structure

d) Plant height

40. **For ripening in banana, ------- is kept along with Ethrel**

a) Sodium Chloride b) Calcium Chloride

c) Sodium Hydroxide d) None of these

41. **Which of the following is a multiple hybrid of Banana**
   a) Bombay Green b) CO.1
   c) Plantain d) None of these

42. **Banana genomic classification was given by**
   a) Simmond and Shepherd
   b) De candolle
   c) Deshmukh and Pandey
   d) Linnaeaus

43. **Botanically banana fruit is a**
   a) Capsule b) Nut
   c) Pome d) Berry

44. **AmritSagar is a variety of**
   a) Banana b) Guava
   c) Fig d) Avocado

45. **A chemical used for artificial ripening in banana**
   a) Ethephon b) Auxin
   c) KNO3 d) Calcium carbide

46. ***Bunchy top virus*was entered into India from**
   a) USA b) Japan
   c) Australia d) Sri Lanka

47. **Choke throat in banana is due to**
   a) High Humidity b) Low Humidity
   c) High temperature d) Low Temperature

48. **Kokkan disease in banana is caused by**
   a) Fungus b) Bacteria
   c) Virus d) MLOs

49. **Major bottleneck in banana breeding is**
   a) Pollen sterility
   b) Non- receptivity of stigma

c) Vegetative parthenocarpy
d) All the Above

50. ***Banana Bract Mosaic Virus* (BBMV) is serious disease in**
a) Poovan b) Rasthali
c) Monthan d) Hill Banana

51. **Seediness in banana can be controlled by the spraying of**
a) NAA @ 25 ppm b) 2,4-D @ 25 ppm
c) Kinetine @ 25 ppm d) ABA @ 25 ppm

52. **Improper bunch filling in banana is due to**
a) K deficiency b) N deficiency
c) B deficiency d) Zn deficiency

53. **Tetrazolium test is used for the diagnosis of presence of**
a) Fungus b) Bacteria
c) Virus d) MLOs

54. **Water logging condition in banana cultivation leads to the more incidence of**
a) Panama wilt b) Sigataka leaf spot
c) Anthracnose d) Mokowilt

55. **Which type of banana bear only female flower buds?**
a) French plantain b) Horn Plantain
c) French horn d) False horn

56. **The edible banana fruit is seedless because of**
a) Embryo abortion
b) Parthenogenesis
c) Vegetative parthonecarpy
d) Stimulativeparthonecarpy

57. **Banana variety which is immune to panama wilt is**
a) Rasthali b) Basarai
c) LalKela d) Poovan

58. ***Banana Bunchy Top Virus*** **is transmitted by**

a) Aphids b) White fly
c) Thirps d) Nematodes

59. Which of the following banana cultivar is used for dual purpose

a) Monthan b) Grand Naine
c) Nendran d) Red Banana

60. The optimum temperature range for banana cultivation is

a) 20 – 30°C b) 10 – 20°C
c) > 40°C d) > 10°C

61. Suitable planting season for banana in subtropical regions is

a) Summer b) Winter
c) Autumn d) Spring

62. In banana, under high density planting (3 suckers/clump) spacing of 1.8 X 3.6 m accommodates ––––––– no. of plants per hectare

a) 4630 b) 3086
c) 2500 d) 2310

63. Most suitable intercrops for banana plantation is

a) Leguminous crops b) Solanaceous crops
c) Cucurbitaceous crops d) None of the above

64. Percentage of yield loss due to weeds in tall cultivars of banana is

a) 90 % b) 45 %
c) 20 % d) 70 %

65. Which of the following banana cultivars is wilt susceptible?

a) Rasthali b) Monthan
c) Virupakshi d) All the above

66. Removal of suckers is one of the critical operations in banana and it's called as

a) Denaveling b) Desuckering
c) Mattocking d) Propping

67. **During bunch covering the polythene bunch sleeve should have ————ventilation**

a) 50 %   b) 10 %
c) 20 %   d) 6 %

68. **Growth regulator recommended for managing Kottavazhai in banana cv.Poovan is**

a) 2,4-D 25 ppm   b) NAA 250 ppm
c) GA3 100 ppm   d) 2,4-D 250 ppm

69. **Commercially cultivated ber species in India is**

a) *Zizuphusrotundifolia*   b) *Z. xylopyrus*
c) *Z. vulgaris*   d) *Z. nummularia*

70. **Botanically ber fruit is**

a) Drupe   b) Nut
c) Pome   d) Berry

71. **Best rootstock species for budding of many ber varieties is**

a) *Zizuphusrotundifolia*   b) *Z. xylopyrus*
c) *Z. rugosa*   d) *Z. mauritiana*

72. **Deciduous type of ber is………………**

a) *Zizuphusrotundifolia*   b) *Z. xylopyrus*
c) *Z. jujube*   d) *Z. nummularia*

73. **Best time for budding in ber is…………..**

a) July - Aug   b) Sept - Oct
c) May - June   d) Feb - March

74. **Ber is commercially propagated by…………..method**

a) Seeds   b) T - budding
c) Inarch grafting   d) Air layering

75. **Salinity and alkalinity tolerant variety of ber is ……………**

a) Gola   b) Umaran
c) Mahatawali   d) Mundia

76. **Ber varieties Katha, Tikdi, Sanaur No.1, Mehrun and Dodia are**
   a) Salinity tolerant
   b) Fruit fly resistant
   c) Alkalinity tolerant
   d) bark eating caterpillar resistant

77. **Which of the following fruit crop is having pruned for flowering?**
   a) Ber b) Litchi
   c) Guava d) Mango

78. **Pollen Sterility is major problem in ………..fruit**
   a) Grape b) Guava
   c) Ber d) Pomegranate

79. **Which of the following fruit has non-endospermic seed?**
   a) Ber b) Beal
   c) Aonla d) Mango

80. **Ganesh Kirti is a variety of**
   a) Grape b) Guava
   c) Ber d) Pomegranate

81. **India is the largest producer of ....in the world.**
   a) Lime b) Sweet orange
   c) Grapefruit d) Mandarin

82. **Mosambi, a popular sweet orange cultivar is predominantly grown in**
   a) Punjab b) Maharashtra
   c) Rajasthan d) Andhra Pradesh

83. **Cross protection technique is effective to check the.................... in lime.**
   a) Tristeza Virus b) Bacterial Canker
   c) Gummosis d) Xoloporosis

84. A salt tolerant rootstock for citrus is

a) *Severeniabuxifolia* b) Trifoliate orange
c) Rough lemon d) Sour orange

85. Among genus *Citrus* which species is most freeze hardy

a) *C. aurantium* b) *C. jambhiri*
c) *C. unshiu* d) *C. limon*

86. Yellow spot, a disorder in citrus is due to deficiency of

a) Mo b) Zn
c) B d) Ca

87. Bacterial canker resistant hybrid of acid lime is

a) Hybrid 2 b) Hybrid 4
c) NS 2 d) All of these

88. Bahar treatment in citrus is generally practiced in ……………

a) East & North India b) Central & West India
c) Central & South India d) West & South India

89. Best rootstock for citrus in South India is

a) Cleopatra mandarin b) Rangpur lime
c) Citrange d) Rough lemon

90. Bitterness in citrus juice is due to presence of……………

a) Acid b) Limonin
c) Glucosides d) Acid: Sugar ratio

91. For Sweet orange var. blood red which is highly suited rootstock in Punjab?

a) Cleopatra mandarin b) Rangpur lime
c) Sour orange d) Rough lemon

92. Canker is most serious problem of – – – – –- citrus fruits

a) Lime b) Sweet orange
c) Grape fruit d) Mandarin

93. **Citranges, the potential rootstock of citrus has originated as a result of cross between**

a) Sweet orange & Trifoliate orange

b) Sweet orange & Mandarin

c) Trifoliate orange & Mandarin

d) Trifoliate orange & Sweet orange

94. ***Citrus myrtifolia*** **is originated as a result of bud mutation from which species of citrus is**

a) *C. aurantium* b) *C. sinesis*

c) *C. reticulata* d) *C. nobilis*

95. **Origin of *Citrus* species is ––––––**

a) South America b) South- East Asia

c) West Indies d) South America

96. **Colour pigment in the flavedo of mature ripe orange fruit is............**

a) Hesperidine b) Naringin

c) Carotene d) Chlorophyll

97. **Common rootstock for budding of mandarin is**

a) *C. aurantium* b) *C. sinesis*

c) *C. limonia* d) *C. jambhiri*

98. **Degreening in citrus is done with application of**

a) Ethephon b) GA3

c) Cytokinin d) Auxin

99. **Flying dragon is the most dwarfing rootstock for..........**

a) Citrus b) Grape

c) Mango d) Apple

100. **Granulation can be reduced by the spraying of.............**

a) Leaf arsenate b) Urea

c) Copper oxide b) KNO3

101. **Growth regulator commercially used for control for fruit drop in citrus is**

a) NAA b) 2, 4 - D

c) IBA d) ABA

102. **Hesperidium is the botanical form of – – – – – – fruit**

a) Citrus b) Grape

c) Mango d) Apple

103. **Indicator plant for *Triteza virus* in citrus is – – – – – – – – – – –**

a) *C. aurantifolium* b) *C. sinesis*

c) *C. reticulata* d) *C. paradisi*

104. **Kinnow is a cross between**

a) *C. nobilis* and *C. deliciosa*

b) C. reticulata and *C. sinesis*

c) *C. reticulate* and *C. sinensis*

d) *C. nobilis* and *C. sinensis*

105. **Monoembryonic species of citrus is**

a) Pummello b) Grapefruit

c) Sweet orange d) Mandarin

106. **On tree storage in citrus is possible with application of**

a) GA_3 b) 2, 4 - D

c) Thio urea d) ABA

107. **Only citrus fruit which contain Malic acid is..............**

a) Kagzi lime b) Sweet lime

c) Sweet orange d) Kinnow

108. **Lemon squash is preserved for a longer period without the discolouration by use of**

a) Malic acid b) Citric acid

c) Benzoic acid d) Potassium meta bisulphite

109. **Rootstock of citrus suited for dry land is**
   a) Karnakhatta b) Jambhiri
   c) Rangpur lime d) C.*maroptera*

110. **Seedlessness in lemon is due to**
   a) Parthenogenesis b) Ovule sterility
   c) Self incompatibility d) Parthenocarpy

111. **The red or pink fleshed seedless grapefruit variety is**
   a) Red blush b) Foster
   c) Ruby d) Thompson

112. **The white fleshed seedless variety of grapefruit is**
   a) Duncan b) Marsh
   c) Triumph d) Saharanpur Special

113. **Vivipary is noticed in which *Citrus* species?**
   a) *C. paradisi* b) *C, sinensis*
   c) *C. grandis* d) *C, reticulata*

114. **Which of the following is chief cause of spreading citrus decline?**
   a) Citrus nematode b) Pest attack
   c) Fungal disease d) Nutritional deficiency

115. **Which of the following sweet orange cultivar is graft incompatible with Trifoliate orange?**
   a) Jaffa b) Valencia
   c) Pineapple d) Mosambi

116. **The most susceptible citrus species for cold injury is**
   a) Lemon b) Lime
   c) Sweet orange d) Mandarin

117. **Which of the following is a serious problem in citrus breeding?**
   a) Self- incompatibility b) Nucellar poly embryony
   c) Viral diseases d) Seedlessness

118. **Red colour in grapefruit juice is due to presence of……….**

a) Carotene
b) Lycopene
c) Xanthophyll
d) Anthocynin

119. **Mottling in citrus is due to**

a) Citrus nematode
b) B deficiency
c) Fungal disease
d) Zn deficiency

120. **The tolerant citrus species for high temperature is**

a) Grapefruit
b) Lime
c) Sweet orange
d) Lemon

121. **Nagpur Orange belongs to ……….group**

a) *Citrus sinensis*
b) *Citrus reticulata*
c) *Citrus deliciosa*
d) *Citrus jambhiri*

122. **'Pineapple' is variety of**

a) *Citrus nobilis*
b) *Citrus reticulata*
c) *Citrus sinensis*
d) *Citrus limon*

123. **Pink coloured pulp is common in**

a) *Citrus limon*
b) *Citrus aurantifolia*
c) *Citrus limettoides*
d) *Citrus paradisi*

124. **An example for nearly thornless lemon cultivar**

a) Assam lemon
b) Baramasi
c) Eureka
d) Lisbon

125. **Citranges are hybrids of**

a) Trifoliate orange and mandarin
b) Trifoliate orange and Kinnow
c) Trifoliate orange and Sweet Orange
d) Sweet Orange and Mandarin orange

126. **A pink fleshed bud sport grapefruit variety is…………..**

a) Duncun
b) Thompson seedlss
c) Triumph
d) Marsh seedless

127. **Wliking' is a kind of**

a) Apple b) Banana
c) Citrus d) Dates

128. **----------- is a variety from Maharastra in acid lime**

a) RagziNimboo b) Pramalini
c) Vikram d) Rasmi

129. **Which of the following crop is/are manmade fruit crop(s)?**

a) Kinnow b) Atemoya
c) Cultivated strawberry d) All of these

130. **Origin of *Citrus reticulata* is**

a) Japan b) China
c) India d) West Indies

131. **Fruits of which of the following group is mostly seedless?**

a) Nagpur Santra b) Khasi mandarin
c) Satsuma mandarin d) King mandarin

132. **The number of Citrus species recognized by Tanaka**

a) 144 b) 44
c) 66 d) 112

133. **Willow leaf mandarin is a parent of**

a) Ponkan Mandarin b) King Mandarin
c) Kinnow Mandarin d) Nagpur santra

134. **Procimequat is a hybrid involving**

a) *Fortunella japonica* b) *Citrus latipes*
c) *Citrus grandis* d) *Citrus sinensis*

135. **--------------- is an ideal rootstock for citrus**

a) Rangpur lime b) Rough lemon
c) Sour orange d) Cleopatra Mandarin

136. **Poor fruit set in Annona is due to**

a) Heterostyly b) Dichogamy
c) Incompatability d) Sterility

137. **Botanical name of Ilama is..................**

a) *Annonasquamosa* b) *Amuricata*
c) *Adiversifolia* d) *Aglabra*

138. **Type of dichogamy observed in** ***Annonamuricata*****is**

a) Protandry b) Protogyny
c) Both of this d) None of these

139. **Annonas are .... fruit**

a) Climacteric b) Non-Climacteric
c) Parthenocarpic d) None of these

140. **Stone fruit formation is a disorder of**

a) Annona b) Mango
c) Wood apple d) Apple

141. **ArkaSahan is a cross between**

a) *A.atemoya* X *A. squamosa*
b) *A.reticulata* X *A.squamosa*
c) *A.reticulata* X *A.squamosa*
d) *A.atemoya* X *A.glabra*

142. **Pond apple is resistant to flooding and damp soil and is used as rootstock for**

a) Apple b) Custard apple
c) Wood apple d) Plum

143. **Among the following Annona species which one is evergreen**

a) Squamosa b) Cherimola
c) *Muricata* d) Reticulate

144. **Which Annona species have largest fruit size**

a) squamosa b) cherimola
c) reticulata d) *muricata*

145. **The finest annona species which can be grown in the tropical lowland is**

a) *A.diversifolia* b) *A.cherimola*

c) *A.reticulata* d) *A.muriculata*

146. **Tetraploid*Annona* sp. is**

a) *A.glabra* b) *A.cherimola*

c) *A.reticulata* d) *A.muriculata*

147. **Red Israel is a natural hybrid between**

a) A.atemoya X A.squamosa

b) *A.cherimola X A.squamosa*

c) *A.reticulata X A.squamosa*

d) *A.atemoya X A.glabra*

148. **Custard Apple Variety APK (Ca) 1 is a**

a) Hybrid b) Mutant

c) Clonal Selection d) Seedling selection

149. **Pollination in annonas is done by**

a) Insects b) Wind

c) Rain d) Birds

150. **Edible part in annonasare**

a) Seed b) Mesocarp

c) Pericarp d) Flesythalmus

151. **……….is commonly used as a rootstock for most of Annonas.**

a) *A.glabra* b) *A. cherimola*

c) *A.reticulata* d) *A. muriculata*

152. **…………state is the leader in Custard Apple production**

a) Andhra Pradesh b) Tamil Nadu

c) Karnataka d) Kerala

153. **In South India, Custard Apple harvesting is done during the month of**

a) August –October b) June- July
c) December- February d) March- May

154. **Custard Apple contains……….sugar.**

a) 10% b) 20%
c) 30% d) 40%

155. **…………is a physiological disorder of annonas**

a) Chicken tongue b) Stone fruit
c) Albinism d) Dry Neck

156. **Datepalm botanical name is………………**

a) Phoenix dactylifera b) *P. humilis*
c) *P. acaulis* d) *P. sylvestris*

157. **Hand pollination is most useful in which of the following fruit crop?**

a) Date palm b) Grape
c) Guava d) Mango

158. **Vegetative propagation of datepalm is through**

a) Runners b) Suckers
c) Slips d) Stolon

159. **Head in fire and foot in water is associated with**

a) Date palm b) Grape
c) Guava d) Mango

160. **Date palm plant bears................ flowers.**

a) Male b) Female
c) Unisexual d) Bisexual

161. **In India, date palm is harvested at................stage.**

a) Dang b) Doka
c) Pind d) All of this

162. **Fig is a..............................species.**

a) Monoecious b) Dioecious
c) Gynodiecious d) Hermaphrodite

163. **Fig is commercially propagated through**

a) Semi hardwood cutting
b) Epicotyl grafting
c) Hardwood cutting
d) Inarch grafting

164. **The individual fruitlet of fig is**

a) Achenes b) Pome
c) Berry d) Nut

165. **Nematode resistant rootstock of fig**

a) *Ficusglomerata* b) *F. bengalensis*
c) *F. religiosa* d) *F. infectotria*

166. **Pollination in fig takes place by *Blastophagapsenses* through**

a) Fasciation b) Parthenogenesis
c) Caprification d) Apomixis

167. **Pollination of the fig flower takes place through**

a) Wasp b) Bee
c) Honey bee d) Bat

168. **In fig, ...onlyproduces pollen.**

a) Poona fig b) Calimryna fig
c) San Pedro fig d) Capri fig

169. **– – – – – – – – is the recently released variety in guava from IIHR, Bangalore**

a) Arka Sahan b) Arka Kiran
c) Hissar Safeda d) Allahabad Safeda

170. **PusaSrijan is a variety of**

a) Mango b) Grapes
c) Guava d) Lemon

171. **In guava, seedlessness is due to**

a) Parthenocarpy b) Steno spermocarpy
c) both a and b d) Polyploidy

172. **———————————— is used as rootstock in Guava to impart dwarfing effect**

a) *Psidiumpumilum* b) *Psidiumcujavillus*
c) *Psidiummolle* d) *Psidiumcattelianum*

173. **The type of pollination happening in guava is —————————————.**

a) Self pollination b) Cross pollination
c) Often cross pollination d) Pathenocarphy

174. **Crop regulation (Bahar treatment) in citrus, pomegranate and guava is achieved through**

a) Withholding of water
b) Root and shoot pruning
c) Deblossoming by application of NAA
d) All the above

175. **Triploid variety of guava is**

a) L-49 b) Seedless guava
c) Allahabad Safeda d) Harijha

176. **Major processed product of ripe guava is**

a) Jam b) Nectar
c) Jelly d) Squash

177. **Behat coconut is an seedless variety of**

a) Litchi b) Grape
c) Guava d) Avocado

178. **Ascorbic acid content in guava is maximum in**
   a) Pulp b) Peel
   c) Seed d) Core

179. **Bronzing in guava is associated with the deficiency of**
   a) Zn b) Mn
   c) Mg d) Cu

180. **Guava bears mostly on**
   a) Current season growth b) One year old Shoot
   c) Spurs d) Very old shoot

181. ***Psidiumfriedrichs thalianum* is used as a rootstock for guava because its imparts**
   a) Dwarfing and resistant to nematode and
   b) Easily propagated
   c) Wilt Graft compatible withal species of *Psidium*
   d) Improves quality and yield

182. **Which of the following fruit is Poor man's apple?**
   a) Guava b) Jackfruit
   c) Mango d) Banana

183. **...............fruit crop is recommended for nutritional security in India**
   a) Guava b) Papaya
   c) Mango d) Banana

184. **Multiple seeded berry is fruit type found in**
   a) Guava b) Jackfruit
   c) Anonna d) Banana

185. **Easiest and cheapest method of propagation in Guava is**
   a) Stooling b) Ground layering
   c) Air layering d) Grafting

186. **Training system followed in guava is**
   a) Open centre b) Central leader
   c) Modified leader d) Spindle bush

187. **Spacing of Guava in meadow orchard system is**

a) 1 x 2 m b) 1.5 x 1.5 m

c) 2 x 3 m d) 2 x 1 m

188. **Highly preferred Bahar treatment in guava is**

a) Ambibahar b) Hasthbahar

c) Mrigahar d) All of these

189. **Mrigbahar is preferred to regulate the.......... season crop**

a) Summer b) Winter

c) Rain d) All of these

190. **TRY (G)-1 is a variety of**

a) Guava b) Jackfruit

c) Grapes d) Banana

191. **National variety of guava is**

a) L-49 b) ArkaMirdula

c) Lalit d) Allahabad Safeda

192. **Anthracnose and fruit canker tolerant guava variety is**

a) Apple colour b) ArkaKiran

c) L-49 d) Allahabad Safeda

193. **Wilt tolerant guava variety is**

a) L-49 b) ArkaMirdula

c) Puas Srijan d) Allahabad Safeda

194. **Behat Coconut is a variety of**

a) Guava b) Jackfruit

c) Grapes d) Coconut

195. **Artocarpin isolated from jackfruit latex is basically a**

a) Protease b) Dehydrogenase

c) Tripernoid d) None of these

196. **Botanically fruit type of jackfruit is**
   a) Sorosis b) Drupe
   c) Syconus d) Stone

197. **Diclinous flowering habit is noticed in**
   a) Date palm b) Jackfruit
   c) Mango d) Passion fruit

198. **Viviparous seed are reported in**
   a) Grapes b) Jackfruit
   c) Mango d) Papaya

199. **Which is the biggest fruit in the world?**
   a) Mango b) Jackfruit
   c) Bread fruit d) Chakotra

200. **National fruit of Bangladesh is**
   a) Mango b) Jackfruit
   c) Bread fruit d) Malaysia

201. **Jackfruit is also known as**
   a) Poor man's apple b) Poor man's food
   c) Poor man's fruit d) None of these

202. **Special fruiting branch lets called 'food stalks' found in which fruit crop**
   a) Date palm b) Jackfruit
   c) Papaya d) Passion fruit

203. **Days required for fruit set to maturity in jack fruit is**
   a) 90-120 days b) 120-140 days
   c) 50-170 days d) 70-100 days

204. **'Jacaline' is extract from jackfruit, which inhibit the growth of**
   a) Cancer b) HIV
   c) Swine fly d) Diabetics

205. **Which of the Jack variety has off season bearing habit?**

a) PLR-1 b) PLR-2
c) PPI-1 d) BURLIAR-1

206. **Mode of pollination in jackfruit is**

a) Wind b) Insects
c) Rain d) Bee

207. **Ploidy level of Jackfruit is**

a) Hexaploid b) Tetraploid
c) Diploid d) Triploid

208. **Among the following *Artocarpus* species,—————— is edible species.**

a) *A. marianensis* b) *A. hirsutus*
c) *A. chaplasha* d) *A. lakoocha*

209. **In Jack, female flower (spike) is having ……… surface and ……… in size than male flower**

a) Smooth, small b) Smooth, large
c) Granular, large d) Granular, small

210. **Sorosis type of fruit is found in…………**

a) Pineapple b) Mulberry
c) Jackfruit d) All of the above

211. **Precocious bearing cultivar of Jackfruit is**

a) Singapore Jack b) Burliar -1
c) Gulabi d) Rasdar

212. **Flattening of spines and thickening of latex in jackfruit are**

a) Disorder b) Maturity Index
c) Diseases d) None

213. **Exotic variety of Jackfruit is**

a) Black Gold b) Lemon Gold
c) Hazari d) All of the above

214. **Muttamvarika is the cultivar of**
   a) Grapefruit b) Jack fruit
   c) Kiwi fruit d) Jamun

215. **The edible part in jack fruit is**
   a) Carpels b) Bracts
   c) Mesocarp d) Endocarp

216. **Which of the following mango cultivar is suitable for pulping**
   a) Alphonso b) Neelum
   c) Sendhura d) Imampasand

217. **Most rapid method of propagation in mango is**
   a) Inarching b) Top working
   c) Bending d) Stone grafting

218. **————————— is the state of India that produces the largest quantity of mango**
   a) Uttar Pradesh b) Madhya Pradesh
   c) Andra Pradesh d) Tamil Nadu

219. **Best combiner parent in most of the hybridization studies in mango is**
   a) Bhahaduran b) Totapuri Red Small
   c) Alphonso d) Neelum

220. **Jelly seed is a physiological disorder of ——————**
   a) Apple b) Mango
   c) Guava d) Grapes

221. **The female parent of PusaLalima is**
   a) Dashehari b) Neelum
   c) Sensation d) Amrapali

222. **Mango malformation is a problem due to ——————**
   a) Virus b) Fungi
   c) Hormonal imbalance d) All the above

223. **Which one of the following fruit crop is not an introduction to India?**
   a) Guava b) Sapota
   c) Mango d) Papaya

224. **– – – – – – is a mango hybrid released by IARI, New Delhi.**
   a) Pusa Peetamber b) Pusa seedless
   c) Pusa Sindhu d) ArkaAruna

225. **The rapid method of propagation in mango is**
   a) Inarching b) Soft wood grafting
   c) Epicotyl grafting d) Veneer grafting

226. **Mango is native of**
   a) Tropical America b) Africa
   c) China d) Indo Burma

227. **The major flowering season of mango is**
   a) Feb–March b) Nov–Jan
   c) April–May d) Aug–Sep

228. **To improve fruiting in "off" season mango, – – – – – should be applied to suppress excess vegetative growth.**
   a) Auxin b) Gibberellin
   c) Cytokinin d) Anti gibberellins

229. **The primary objective of pruning in mango is**
   a) Photosynthesis b) Sunlight interception
   c) Disease control d) Reducing canopy

230. **Pruning may be done in mango during**
   a) April - May b) Jan - Feb
   c) Aug - Sep d) Nov - Dec

231. **The scientific method of ripening in mango is by using**
   a) Calcium carbide b) Potassium nitrate
   c) Ethrel+NA OH pellets d) Gibberellic acid

232. **Alphonso mango is a variety originated from**

a) Rajapalayam b) Uttar Pradesh
c) Sanga Reddy d) Ratnagiri (Maharastra)

233. **The partents of Hybrid Sindhu mango is**

a) Neelum x Dashehari b) Neelum x Alphonso
c) Ratna x Alphonso d) Alphonso x Banganapalli

234. **A late-season mango cultivar grown in south India is..........**

a) Alphonso b) Ratna
c) Neelum d) Dashehari

235. **Spongy tissue is a physiological disorder in -------- mango variety**

a) Alphonso b) Ratna
c) Neelum d) Dashehari

236. **The commercial method of mango propagation is**

a) Rooting of Hardwood cuttings
b) Soft wood grafting
c) Air Layering
d) T-budding

237. **Among TNAU varieties of Mango, the following variety is dwarf in stature and suitable for high density planting.**

a) Neelum b) PKM-1
c) PKM-2 d) Paiyur - 1

238. **The rootstock used to impart dwarfism in Alphonso Mango is**

a) Mulgoa b) Rumani
c) Vellaikolumban d) Sindhu

239. **The chemical used to induce off-season flowering & to overcome alternate bearing in Mango is**

a) NAA b) Paclobutrazol
c) Etheral d) ABA

240. **The mango variety suitable for high density planting in North India is**

a) Amarapalli b) Co-1

c) Langra d) Mallika

241. **The chemical used to reduce malformation in Mango is**

a) NAA @ 200 ppm b) GA @1000 ppm

c) ABA @ 25 ppm d) Zeatin @ 5000ppm

242. **Example for climacteric fruit is**

a) Mango b) Grape

c) Lemon d) Pineapple

243. **Which variety of mango is preferred in export market**

a) Neelum b) Banglora

c) Alphonso d) Rumani

244. **Which of the following is polyembryonic variety of mango?**

a) Bangalora b) Olour

c) Naducholai d) Khader

245. **Which of the following mango variety is a biennial bearer**

a) Swarnarekha b) Neelum

c) Bangalora d) None of these

246. **------------- hybrid of mango is referred as seedless mango**

a) Ratna b) Manjiri

c) Mallika d) Sindhu

247. **Type of parthenocarpy in Sindhu variety of mango is ?**

a) Stimulative b) Stenospermocarpy

c) Vegetative d) None of these

248. **Alphonso a variety of mango is mostly grown in**

a) Punjab b) UP

c) Maharashtra d) Bihar

249. **Early maturing mango variety of south India is**
   a) Neelum b) Dashehari
   c) Alphonso d) Banganapalli

250. **Which of the following is an off season variety of mango in north India ?**
   a) Mallika c) Fazli
   c) Niranjan d) Olour

251. **Ideal variety for long distance transporting in mango is**
   a) Neelum b) Langra
   c) Bombay green d) Rumani

252. **– – – – – – is a mango hybrid released by IARI, New Delhi.**
   a) Mallika b) Rathna
   c) Sindhu d) Alphonso

253. **Alphonsois the one of the parents for which of the following mango hybrid?**
   a) ArkaAruna b) Pusa Lalima
   c) PusaPeetamer d) Pusa Giant

254. **Sunlight interception is the primary aim in mango for**
   a) Training b) Pruning
   c) Top working d) Smudging

255. **Jelly seed is a physiological disorder in – – – – – – – – – –.**
   a) Apple b) Mango
   c) Guava d) Grapes

256. **In which group will you classify mango according to its ethylene production rate?**
   a) Very low b) Low
   c) Moderate d) High

257. **Sap burn is a postharvest disorder in – – – – – – – – –**
   a) Mango b) Apple
   c) Orange d) Grape

258. **Among fruits, the highest vitamin A content is in mango. It is about**
   a) 200 IU/100g b) 1500 IU/100g
   c) 4000 IU/100g d) 3000 IU/100g

259. **Mango hybrid Mallika is a cross between**
   a) Dashehari X Langra b) Neelum X Dashehari
   c) Neelum X Langra d) Banganapalli X Alphonso

260. **Spongy tissue, a serious problem in mango is seen in**
   a) Totapuri b) Malda
   c) Alphonso d) Langra

261. **Tapka is associated with maturity of**
   a) Litchi b) Aonla
   c) Pineapple d) Mango

262. **A common pest during flowering time of mango (Feb-March) is**
   a) Red ants b) Shoot borer
   c) Hoppers d) Mealy bug

263. **A disease which can transform an "On year" of fruiting to "Off year" in mango is**
   a) Powdery mildew b) Anthracnose
   c) Leaf blight d) Stem canker

264. **A serious physiological disorder of mango orchards near brick kiln is**
   a) Black tip b) Black rot
   c) Spongy tissue d) Malformation

265. **Amimgola is a variety of**
   a) Litchi b) Guava
   c) Fig d) Mango

266. **Black tip of mango is also known as**
   a) Chimney disease b) Soft nose of mango
   c) Witch's broom d) Seed jelly of mango

267. **Creeper is a variety of**

a) Litchi b) Kiwifruit
c) Grape d) Mango

268. **Dwarf mango suitable for kitchen garden is**

a) Amrapali b) Alphonso
c) Mallika d) Dashehari

269. **Flowering in mango can be suppressed by the application of**

a) Auxin b) Gibberellins
c) Kinetin d) Paclobutrazol

270. **Hot water treatment for control of anthracnose in mango is at**

a) 52°C b) 32°C
c) 70°C d) 100°C

271. ***In situ* method of mango grafting is**

a) Soft wood grafting b) Inarching
c) Epicotyl grafting d) Stooling

272. **Mango inflorescence contains which type of flowers**

a) Male and hermaphrodite
b) Male, female and hermaphrodite
c) Male and neutral
d) Female and hermaphrodite

273. **Mango is stored at temperature (°C) of**

a) 6-7 b) 7-8
c) 8-9 d) 9-10

274. **Mango seed lose their viability within**

a) 3-4 week b) 4-5 week
c) 5-6 week d) 6-7 week

275. **Mango variety resistant to mango malformation is**

a) Bhadauran b) Chausa
c) Dashehari d) Landra

276. **Most ideal method for top working in mango is**

a) Soft wood grafting b) Inarching
c) Epicotyl grafting d) Side grafting

277. **Off season mango variety is**

a) Neeranjan b) Alphonso
c) Bombay green d) Dashehari

278. **Only allopolyploidy variety reported in mango is**

a) Vellaikulamban b) Olour
c) Bombay green d) Kurrukan

279. **Pre harvest fruit drop in mango can be controlled by spraying of**

a) NAA @ 100 ppm b) 2,4-D @ 20 ppm
c) IBA @ 20 ppm d) GA3 @ 20 ppm

280. **Rapid and convenient method of grafting in mango is**

a) Soft wood grafting b) Inarching
c) Epicotyl grafting d) Side grafting

281. **Sindhu, a variety of mango has been developed as a result of back cross between**

a) Alphonso X Banganapalli
b) Ratna X Neelum
c) Neelum X Langra
d) Ratna X Alphonso

282. **Specific gravity used as maturity indices for mango is**

a) 1.01 – 1.02 b) 1.5 – 2.0
c) 2.01 – 2.02 d) 2.0 – 2.5

283. **The main insect reported to be responsible for pollination in mango is**

a) Common house fly b) Honey bee
c) A dipteroud fly d) Haver fly

284. **Type of incompatibility in mango is**
   a) Sporophytic b) Gametophytic
   c) Both of this d) None of these

285. **Which country has the highest productivity of mango?**
   a) China b) Venzuela
   c) India d) Brazil

286. **Which of the following mango hybrid is dwarf?**
   a) ArkaAruna b) Ratna
   c) ArkaAnmol d) Arkapuneet

287. **Which of the following mango varieties are regular bearing?**
   a) Bangalora, Neelum, Rumani
   b) Alphonso, Mallika, Amarapali
   c) Dashehari and Langra
   d) Dashehari, Bombay green

288. **Which type of polyploidy occurs in mango?**
   a) Amphidiploidy b) Autopolyploidy
   c) None of these d) Both of this

289. **'Hafus' mango is also known as**
   a) Neelum b) Alphonso
   c) Langra d) Dashehari

290. **Polyembryonic salt resistant rootstock of mango is**
   a) Kurrukan b) Vellakolumban
   c) Langra d) Olour

291. **Black tip in mango is controlled by**
   a) Caustic soda and lime spray
   b) Fungicide spray
   c) Nutrient spray
   d) Pesticide spray

292. **Ornamental mango varieties are**

a) Neelum, Rumani b) Croton and ChitlaAfag
c) Malda and Olour d) Dashehari, Bombay green

293. **Method of producing homozygous rootstock in mango is**

a) Seeds b) Grafting/budding
c) Cuttings/Stooling d) Layering

294. **The HQ of International Mango registration Society is at**

a) IARI, New Delhi b) IIHR, Bangalore
c) CISH, Lucknow d) TNAU, Coimbatore

295. **Which of the following variety of mango have irregular bearing habit ?**

a) Bangalora b) Neelum
c) Amrapali d) Dashehari

296. **Which of the following fruits even though propagated by seed do not have variability**

a) Litchi b) Mangosteen
c) Loquat d) Papaya

297. **Which of the following pair is not correctly matched?**

a) Datepalm-Monoecious b) Papaya-Polygamous
c) Walnut-Monoecious d) Pistachionut–Dioecious

298. **A transgenic papaya variety is**

a) Maradol b) Red Lady
c) Sunset Solo d) Sun Up

299. **Scientist associated with studies on sex forms in papaya is**

a) Drew b) N.W.Simmonds
c) W.B.Storey d) R.W.Allard

300. **Dioecious mutant variety of papaya is**

a) CO.8 b) Washington
c) Pusa Giant d) PusaNanha

301. **Papain is – – –- in water**

a) Insoluble b) Highly soluble

c) Moderately d) Less soluble

302. **– – – – – – – – – is a transgenic papaya variety developed against *Papaya Ring spot Virus***

a) Surya b) Rainbow

c) PusaNanha d) Red lady

303. **– – – – – – – – – – – –- is added as preservative to wet latex from papaya before drying**

a) Potassium chloride

b) Sodium chloride

c) Potassium meta bisulphite

d) Potassium dihydrogen phosphate

304. **CO. 1 papaya is derived from,**

a) Ranchi b) Washington

c) Sun Rise Solo d) Coorg Honey Dew

305. **The seed rate for papaya for one hectare is**

a) 250 g b) 500 g

c) 625 g d) 1000 g

306. **The most devastating disease in papaya is**

a) Foot rot b) Collar rot

c) Papaya ring spot virus d) Leaf spot

307. **Cold tolerant papaya species is known as – – – – – – – – – – – – – – –.**

a) *Caricacandamarcensis* b) *Carica papaya*

c) *Caricacauliflora* d) *Caricagoudotiana*

308. **CO 3 and CO 7 are examples of – – – – – – – – – – – –- varieties of papaya**

a) Dioecious b) Monoecious

c) Gynodioecious d) Andromonoecious

309. **Femaleness in papaya is controlled by genotype**
   a) M1M b) M2M
   c) M1M2 d) mm

310. **CO.8 papaya variety is a cross between**
   a) CO.2 X Sunrise Solo
   b) Sunrise Solo X Pink Flesh Sweet
   c) CO.2 X Red Anthered Male parent (CO-2)
   d) CO.1 X Washington

311. **Transgenic PRSV resistant variety is**
   a) Sunrise Solo b) Sunup
   c) Rainbow d) b & c

312. **High carotene content papaya variety is**
   a) Sunrise Solo b) CO.8
   c) Rainbow d) Red Lady

313. **Based on the climatic requirement, papaya is classified as a**
   a) Temperate fruit b) Tropical fruit
   c) Sub-tropical fruit d) Sub temperate fruit

314. **Genotype of the sex reversing male papaya is**
   a) M1MRR& M1M b) M2MRR
   c) M1M2 d) Mm

315. **Monoecious species of papaya is**
   a) *V.asconcellea*monoica b) *V. candamorcensis*
   c) *V. pentagona* d) *V. gudotiana*

316. **Origin of papaya is**
   a) Tropical America b) Temperate America
   c) China d) India

317. **Outer layer of papaya seed which hinders germination is called as**
   a) Integument b) Testa
   c) Sarcotesta d) Embryo

318. **The maximum papain yielding papaya selection is**
   a) Washington Honey Dew
   b) CO 5
   c) CO 2
   d) CO 6

319. **Vitamin A content in papaya is**
   a) 2020 IU/100g b) 1020 IU/100g
   c) 4000 IU/100g d) 3000 IU/100g

320. **Resistance source of mosaic/ leaf curl virus in papaya breeding is**
   a) *V.asconcelleamonoica* b) *V. candamorcensis*
   c) *V. cauliflora* d) *V. gudotiana*

321. **Singapore, Ceylon, Madhubindu are the varieties of**
   a) Litchi b) Pineapple
   c) Papaya d) Mangosteen

322. **The most susceptible crop to water logging condition is**
   a) Litchi b) Banana
   c) Papaya d) Mangosteen

323. **The pigment responsible for coloration in papaya is**
   a) Chlorophyll b) Carotinoids
   c) Anthocynin d) Caricaxanthin

324. **Which of the following is a frost resistant species of papaya?**
   1. *V.asconcellea*monoica 2. *V. candamorcensis*
   3. *V. cauliflora* 4. *C. papaya*
   a) 1 and 2 b) 2 and 3
   c) 1,2,3 d) 2

325. **The sex form more stable in papaya is**
   a) Male b) Female
   c) Hermaphrodite d) None of these

326. The most limiting factor in papaya cultivation in North India is

a) Temperature b) Rain
c) Wind d) Frost

327. Which state of India stands 1st position in area and production of papaya?

a) Andhra Pradesh b) Tamil Nadu
c) Kerala d) Gujarat

328. The storage temperature in papaya for export market is

a) 7-9°C b) 14-15°C
c) 10-13°C d) 6-7°C

329. Among, which species of papaya is resistant to frost

a) *C. pubescence* b) *C. monoica*
c) *C. pentagona* d) *C. cauliflora*

330. Among the following varieties, which variety suitable for dual purposes (Dessert purpose andpapain extraction)................?

a) CO.3 b) CO.2
c) CO.4 d) CO.5

331. The only red flesh cultivar in dioecious type of papaya is

a) PusaNanha b) CO.3
c) Solo d) CO-8

332. Popular IIHR variety with good peeling quality

a) Arkasurya b) Solo
c) Washington d) Ranchi

333. Proteolytic enzyme present in papaya is..............

a) Bromeline b) Caricaxanthin
c) Papain d) none of these

334. The gene responsible for tolerance to PRSV in transgenic papaya is

a) PRSV CP gene b) PRV CP transgene
c) CP gene d) PRV gene

335. **Exotic papaya cultivar tolerant to PRSV is**
   a) Taiwan b) Solo
   c) Red lady d) Eksotica

336. **The seed rate for gynodioecious cultivar of papaya (g/ha) is**
   a) 200-250 b) 300
   c) 400 d) 500

337. **The ratio of male and female in dioecious line is maintained by**
   a) 1:20 b) 1:10
   c) 1:15 d) 1:8

338. **Which of the following fruit crop is called as 'tree melon'?**
   a) Papaya b) Jackfruit
   c) Mango d) Durian

339. **Example for continuous flowering and fruiting crop is**
   a) Papaya b) Sapota
   c) Mango d) Banana

340. **Sib mating is practiced in**
   a) Papaya b) Banana
   c) Mango d) Guava

341. **Example for monoecious papaya species is............**
   a) *V.asconcellea*monoica b) *V. candamorcensis*
   c) *V.cauliflora* d) *V. gudotiana*

342. **Red pulpcolour is present only in ....papaya**
   a) Monoecious b) *Dioecious*
   c) *Gynodioecious* d) None of these

343. **How many species in the genus *Carica* ?**
   a) 2 b) 4
   c) 1 d) 3

344. **Jalore seedless, a pome granate variety is recommended for cultivation in the state of**

a) Rajasthan b) U.P
c) Gujarat d) Maharashtra

345. **Jyothi pomegranate is a clonal selection from**

a) Kandhari b) Alandi
c) Ganesh d) Bassein seedless

346. **Mirdula is a hybrid of**

a) Pomegranate b) Guava
c) Litchi d) Sapota

347. **Origin of pomegranate is**

a) India b) Italy
c) Irag d) Iran

348. **Pomegranate fruit is botanically**

a) Sorosis b) Pome
c) Balusta d) Stone

349. **Pomegranate is preferably trained to**

a) Single stem system b) Multiple stems system
c) Open centre d) Central leader

350. **PKM-1, a popular variety of sapota is a selection from**

a) Guthi b) Kirthibarthi
c) Thagarampudi d) Dwarapudi

351. **-------------- problem is overcome by mixed planting of varieties in sapota**

a) Parthenocarpy b) Protandary
c) Protogyny d) Stamen carpellody

352. **---------- is the most ideal rootstock for sapota**

a) Pala b) Chiku
c) Mahua d) Bassia

353. **––––––––––––––– is the commercial method of propagation in Sapota**
   a) Soft wood grafting b) Veneer grafting
   c) Stone grafting d) Inarch grafting

354. **Chicklet is obtained from the latex of**
   a) Sapota b) Pomegranate
   c) Papaya d) Mango

355. **Cock's comb is a physiological disorder of**
   a) Sapota b) Plum
   c) Pear d) Apple

356. **Cross pollination in sapota is due to**
   a) Protogyny b) Homogamy
   c) Protandry d) None of these

357. **Orchard decline is major problem in**
   a) Sapota b) Pomegranate
   c) Papaya d) Mango

358. **Example for Spindle shape sapota variety is**
   a) PKM.1 b) PKM.2
   c) PKM. 4 d) PKM. 5

359. **Off season bearing variety of Sapota is**
   a) PKM.1 b) PKM.2
   c) PKM.4 d) PKM. 5

360. **Which of the following training system is followed in sapota**
   a) Open centre b) Central leader
   c) Modified leader d) Spindle bush

361. **Which species of *Manilkara* is known as Adams' apple ?**
   a) *M. hexandra* b) *M. kauki*
   c) *M. bilerata* d) none

362. **In Sapota, a cross between Oval X Round shape gives ....shape**

a) Spinddle b) Oval
c) Round d) Oblong

363. **Edible part of Sapota is**

a) Placenta b) Mesocarp
c) Endocarp d) Endosperm

364. **Chickle is extracted from...........parts of sapota**

a) Root, stem b) Bark, immature fruit
c) Ripen fruit, seed d) none

365. **Flattening of branches in Sapota is due to**

a) Fungus b) Bacteria
c) Virus d) MLO

366. **Round shape variety of Sapota is**

a) CO1 b) PKM2
c) CO3 d) PKM1

367. **Central Institute for Arid Horticulture (CIAH) is located at ___________.**

a) Jodhpur b) Abohar
c) Bikaner d) Hissar

368. **The head quarters of Bioversity International is located at ____________.**

a) Nigeria b) Germany
c) France d) Italy

369. **International year of biodiversity as per declaration of UN ....**

a) 1997 b) 2009
c) 2010 d) 2011

370. **Which is the cheapest preservation method of fruits?**
   a) Canning b) Freezing
   c) Drying d) Fermentation

371. **In which form, pectin is found in fruit?**
   a) Pectin b) Calcium pectate
   c) Proto-pectin d) None of these

372. **During controlled atmospheric storage, composition of which of the following set of gases is controlled?**
   a) O_2+N_2 b) CO_2+N_2
   c) C_2H_4+N_2 d) CO_2+O_2

373. **In precooling field heat is mostly removed by ------.**
   a) Conduction b) Convection
   c) Radiation d) None of these

374. **Degreening of fruits is caused by ---------.**
   a) Loss of chlorophyll b) Loss of lycopene
   c) Loss of anthocyanin d) Loss of carotene

375. **The enzyme which is responsible for softening of fruits**
   a) Poly sulphonase b) Catalase
   c) Polyhydrase d) Polygalactouranase

376. **----------- hormone helps in improving fruit set in fruit trees**
   a) Ethylene b) Abscissic acid
   c) Salicylic acid d) NAA

377. **Which of the following chemical is used as Cryoprotectant**
   a) DMSO b) 2, 4-D
   c) Ethylene d) EMS

378. **Enzyme responsible for formation of ethylene is -----.**
   a) ACC synthase b) ACC transferase
   c) ACC Oxidase d) Adenosyl methionine

379. **The fruit product order (FPO) license is given by ---------**

a) GOI b) NHB
c) ICAR d) State Government

380. **As fruits mature, the specific gravity will ---------.**

a) Increase b) Decrease
c) Remains constant d) None of these

381. **In India, processing industries utilize ....% of total fruit production?**

a) >1% b) <1%
c) 2-3% d) 5%

382. **Which of the following is a rapid precooling method?**

a) Forced air cooling b) Hydro cooling
c) Vacuum cooling d) Evaporate cooling

383. **Calcium carbide releases --------.**

a) Carbondioxide b) Ethylene
c) Acetylene d) Methane

384. **SINGER is related to __________.**

a) Field gene bank
b) *In vitro* conservation
c) Data base management
d) Cryo preservation

385. **The aroma in fruit is caused due to the formation of ----------.**

a) Fats b) Essential oil
c) Esters d) Aromatic compounds

386. **Apomixis is the phenomenon where ----------- develop into embryo**

a) Nucellar tissues b) Endosperm
c) Nucleus d) Zygote

387. **--------- is one of the methods of eliminating dormancy where the seeds are subjected to chilling temperature.**

a) Scarification b) Stratification
c) Vernalisation d) Seed coat removal

388. **Use of------------- is advised to avoid incomp atibility problems where incompatible partnersare grafted.**

a) Hormones b) Interstocks
c) Paraffin wax d) Clips

389. **--------- is the principle that is the basis for successful micropropagation.**

a) Somatic embryogenesis b) Organogenesis
c) Totipotency d) Hardening

390. **Climacteric fruit is one where -------- reaches a climacteric peak during the course of ripening.**

a) Colour change b) Sugar accumulation
c) Respiration rate d) Transpiration

391. **Alternate bearing is a problem caused by --------.**

a) Nutritional deficiency
b) Unbalanced growth of vegetative and reproductive parts
c) Fungal infestation
d) Water deficit

392. **------------- is a problem in mutation breeding.**

a) Stability b) Availability of chemicals
c) Human safety d) Response

393. **----------- is the immobile nutrient.**

a) N b) P
c) K d) Mg

394. **Primary objective of training in fruit crops is to impart...........**

a) Frame work
b) Reduce number of shoots

c) Better light interception
d) Reduce plant height

395. ________ **nutrient is required at higher quantity in fruit crops for quality enhancement.**

a) N b) P
c) K d) Fe

396. **In fertigation technology ---------- is the important consideration based on which, scheduling of nutrients is planned.**

a) Plant population b) Plant height
c) Nutrient requirement d) Nutrient source

397. **The maximum loss of perishable produce happens during transit due to ---------.**

a) Untimely harvest b) Bruising damage
c) Improper storage d) All the above

398. **------- is the important certificate needed for export of horticulture produce**

a) HACCP b) Sanitary &Phytosanitary
c) Organic certification d) GAP

399. **----------- is the most important activity expected in plant system after layering.**

a) Rooting b) Heeling
c) Shooting d) Flowering

400. **CISH, is located at ____________.**

a) Bangalore b) Trichy
c) Lucknow d) Srinagar

401. **The nucellar embryos develop through the process of_________.**

a) Poly embryony b) Bud mutation
c) Parthenocarphy d) Apomixis

402. **———————— hormone helps in improving fruit set in fruit trees.**

a) Ethylene b) Abscissic acid
c) Salicylic acid d) NAA

403. **Cryo preservation is associated with ————————.**

a) Liquid Nitrogen b) Liquid oxygen
c) Liquid Carbon dioxide d) None of the above

404. **Indian Institute of Horticultural Research (IIHR. is located at —————.**

a) Lucknow b) New Delhi
c) Bangaluru d) Ranchi

405. **———————— is one of the popular hormones used for root induction.**

a) GA3 b) Cytokinin
c) IBA d) ABA

406. **————————- is the method of grafting by which old unproductive orchards are regenerated**

a) Top working b) Bridge grafting
c) Epicotyl grafting d) In situ grafting

407. **Chilling requirement means prevalence of low temperature conditions for a duration in case of ————————- crops.**

a) Humid zone b) Subtropical
c) Temperate d) Aridzone

408. **———————— is the hormone which promotes vegetative growth thereby delaying flowering**

a) Auxin b) Gibberellin
c) Cytokinin d) Abscissic acid

409. **Alternate bearing is a problem caused by ———————-**

a) Nutritional deficiency
b) Unbalanced growth of vegetative

c) Fungal infestationanreproductive parts
d) Water deficit

410. **--------- is the basic concept in organic farming.**
a) LEISA b) IFOAM
c) EUREPGAP d) HACCP

411. **-----------nutrient is required at higher quality in fruit crops for quality enhancement**
a) N b) P
c) K d) Fe

412. **Under high density planting, -------- is an important requirement.**
a) Gap filling
b) Weeding
c) Nutrition at higher doses
d) Application of growth regulator

413. **Pollinizers are essential in --------- fruit crops for good fruitfulness**
a) Tropical b) Subtropical
c) Temperate d) Arid

414. **--------- is the recent and improved method of storage for fruits and vegetable.**
a) Cold storage b) MA storage
c) Ambient conditions d) Vacuum packing

415. **By waxing -------- process is showed down in fruits and vegetable during storage**
a) Ripening b) Respiration
c) Transpiration d) Photosynthesis

416. **Which refrigerant is commonly used in cold storage in India?**
a) Ethylene b) Calcium carbide
c) Ammonia d) Sodium benzoate

417. **Chemical used for removing the spray residues of Arsenic and Lead from the fruit is – – – – –**

a) NaOH b) HCl
c) NaCl d) H2SO4

418. **Alcholol content of wine ranges from – – – – – – –.**

a) 5-10% b) 10-12%
c) 7-20% d) 12-17%

419. **Which of the following is a rapid precooling method?**

a) Forced air cooling b) Hydro cooling
c) Vacuum cooling d) Evaporate cooling

420. **Amount of pectin in jelly is tested by – – – – – –.**

a) Penetrometer b) Refractometer
c) Thermometer d) Jelmeter

421. **– – – – – – – – – is the recent method followed in citrus for production of disease free planting materials.**

a) Callus culture b) Micro grafting
c) In situ grafting d) Shoot top culture

422. **The most common method for grape preservation is – – – –.**

a) Jam making b) Jelly making
c) Dehydration d) Pickling

423. **Stenospermocarpy is expressed in**

a) Italia b) Tas-A-Ganesh
c) Red Globe d) Anab-E-Shahi

424. **Plant growth regulator used to increase berry size in grape is – – – – – – – – –**

a) IAA b) GA3
c) Cytokinin d) ABA

425. **Among the following which is false**

a) Most of the *vinifera*grapes are self fertile
b) Crosses between different *Euvitis*species can

c) Parents withhigh seed index are preferablefor breeding seedless varieties be easily made

d) *Euvitis*and *Muscadina* hybrids are fertile and vigourous

426. **In grapes, fruitbuds appear in ---------- which is a past season shoot.**

a) Shoot b) Spur

c) Arm d) Cane

427. **--------- is the root stock which has good to lerance against nematodes and drought in grapes.**

a) Dog Ridge b) 16-1

c) Khirnee d) Vellaikolumban

428. **Saline tolerant root stock in grape is __________**

a) Salt Creek b) St. George

c) 1613 d) Rupestnis Du Hot

429. **NRC grape is located at ___________**

a) Nagpur b) Dehradun

c) Hyderabad d) Pune.

430. **Development of seedless varieties is possible in grape due to -------------.**

a) Triploid breeding b) Stenospermocarpy

c) Parthenocarpy d) Hormones

431. **----------- is a mutant of grape derived from Thompson seedless**

a) Sharad seedless b) Sonaka

c) PusaUrvasi d) Tas-e-Ganesh

432. **The richest source of iron is __________**

a) Dry karonda b) Date palm

c) Bael d) Woodapple

433. **Patch budding is generally practiced in ---------.**

a) Guava b) Sapota
c) Aonla d) Ber

434. **Which of the following belongs to family Rutaceae?**
a) Wood apple b) Mulberry
c) Jamun d) Ber

435. **Heat units required for flowering, fruit development and ripening in date palm is – – – – –**
a) 3300 b) 4000
c) 2250 d) 1100

436. **Ganesh variety of pomegranate is selection from – – – – –.**
a) Kandhari b) Alandi
c) Dholka d) Nabha

437. **Term 'bahar' is related to the cultivation of – – – – – – – – –.**
a) *Musa paradisiacal* b) *Litchi chinensis*
c) *Maluspumila* d) *Punicagranatum*

438. **The fruit of pomegranate is botanically known as**
a) Balusta b) Berry
c) Drupe d) Multiple fruit

439. **Fruit of ber is – – – – – – – –**
a) Drupe b) Berry
c) Balusta d) Pome

440. **In Tamil Nadu, pruning in ber should be done during – – – – – – – –.**
a) February-March b) December-January
c) April-May d) January-July

441. **Which one of the following is the late ripening cultivar of ber?**
a) Umran b) Gola
c) Kaithali d) Seb

442. **Dwarfing rootstock used in ber is – – – – – –.**

a) *Z.nummularia* b) *Z.lotus*

c) *Z.rotundifolia* d) *Z.jujube*

443. **Which of the following is the most suitable for growing under arid zones of India?**

a) Papaya b) Jack

c) Sweet orange d) Ber

444. **Fruit necrosis in aonla is due to– – – – – – – –.**

a) B deficiency b) Ca deficiency

c) K deficiency d) Zn deficiency

445. **– – – – – – – – – -group of pine apple is highly suitable for canning.**

a) Queen b) Kew

b) Yellow Mauritius d) Red Mauritius

446. **The planting system highly successful in pineapple is _________.**

a) Hedge system

b) Double hedge system

c) Paired row planting system

d) None of these

447. **Pineapple takes – – – – – – – – – – months from flowering to ripening.**

a) 6-7 months b) 3 ½ to 4 months

c) 6 ½ to 7 ½ months d) ½ to 5 ½ months

448. **Dry neck is a physiological disorder**

a) Fig b) Avocado

c) Cherries d) Persimmon

449. **Type of fruit in Avocado is**

a) Drupe b) Nut

c) Berry d) Achene

450. **Fuerte is a cross between __________**

a) Mexican and Guatamalan race
b) Guatamalan and West Indian race
c) Mexican and West Indian race
d) West Indian and Guatamalan race

451. **Hydroxy citric acid is found in**

a) Carambola b) Rambutan
c) Kokam d) Litchi

452. **Which of the following fruit is a rich source of Riboflavin**

a) Jamun b) Aonla
c) Bael d) Wood apple

453. **Yellow passion fruit is botanically**

a) *Passifloraligularis*
b) *Passifloraedulisvar.flavicarpa*
c) *Passifloracaerulea*
d) *Passifloraedulis*

454. ________ **is a F1 hybrid developed by IIHR in passion fruit**

a) San Roman b) Kauveri
c) ArkaAnmol d) ArkaShyam

455. **Forgood bud-break in apple — — — — — — — — is recommended.**

a) Sandolin A b) GA
c) NAA d) MH

456. **— — — — -- is the apple rootstock imparting dwarfness.**

a) Crab Apple b) Robusta
c) Northern Spy d) M 27

457. **Browning in apple is due to the enzyme — — — — — — — — — —.**

a) Hydrogenase b) Polyphenol oxidase
c) Polygalactouranase d) Peroxidase

458. **In which group will you classify apple according to its respiration rate?**

a) Low
b) Very low
c) Moderate
d) High

459. **Spur type of apple root stock**

a) MM 116
b) MM 106
c) MM 109
d) MM 110

460. **A single sigmoid growth curve is noticed in ___________.**

a) Grapes
b) Peach
c) Guava
d) Apple

461. **In apple, colour development is more intensive in __________.**

a) Lower hills
b) Mid hills
c) High hills
d) Not affected by Altitude

462. **Pollinizer for Red Delicious apple is**

a) Jonathan
b) Ambri
c) Early Shanburry
d) Golden Delicious

463. **Optimum temperature for the stratification of apple seed is _____ .**

a) 8°C
b) 4°C
c) 0°C
d) 10°C

464. **The apple trees require pollinizer varieties since they are — — — — — — — — — — —**

a) Parthenocarpic
b) Cross incompatible
c) Self incompatible
d) Sterile

465. **Under heavy soils and excessive soil moisture conditions, apricot grows better on which rootstock?**

a) Wild apricot
b) Wild peach
c) Myrobalan plum
d) Quince

466. **Phalsa is native to – – – – – .**
   a) India b) China
   c) Brazil d) Mediterranean region

467. **The most suitable fruit crop for a soil of 9.5 pH is**
   a) Jamun b) Jack
   c) Phalsa d) Loquat

468. **Among the different fig types, the one which produces viable pollen is**
   a) Caprifig b) Smyrna fig
   c) San Pedro d) Common fig

469. **Popularly grown cultivar of Fig in India is __________.**
   a) Conadria b) Excel
   c) Poona d) Brown Turkey

470. **Stella, a variety of cherry has originated as a result of**
   a) Chance Seedling b) Hybridization
   c) Induced mutation d) Polyploidy

471. **Fruit which is not peeled before canning is – – – – – – – – – .**
   a) Pineapple b) Mango
   c) Cherry d) Apple

472. **Prunes are prepared from – – – – – – – .**
   a) Peach b) Plum
   c) Pear d) Apple

473. ***Syzygiumjambos* is known as _________.**
   a) Malay apple b) Velvette apple
   c) Rose apple d) None of the above

474. **Heavy fruit drop in Jamun can be controlled by spraying _____.**
   a) 60 ppm of GA3 twice one at full bloom and other 15 days after fruit set
   b) 100 ppm of GA3 twiceone at full bloom and other 15 days after fruit set

c) 150 ppm of GA3 twice one at full bloom and other 15 days after fruit set

d) 100 ppm of NAA twice one at full bloom and other 15 days after fruit set

475. **Zespri Gold is a variety of**

a) Grapes
b) Kiwi fruit
c) Sweet Cherry
d) Quince

476. **------ is the pear variety suitable for South Indian hills like Kodaikanal.**

a) Bartlett
b) Gola
c) Anjou
d) Jargonelle

477. **Heterostyly and incompatibility is found in __________.**

a) Mangosteen
b) Litchi
c) Carambola
d) Loquat

478. **Which of the following is a member of Sapindaceae**

a) Sapota
b) Jack
c) Litchi
d) Mangosteen

479. **Matured litchi fruit is botanically called as __________.**

a) Multiple fruit
b) Berry
c) Drupe
d) Nut

480. **Metaxenia is expressed in __________.**

a) Litchi
b) Apple
c) Persimmon
d) Rambutan

481. **A fruit cracking resistant cultivar of litchi**

a) Deshi
b) China
c) Rose scented
d) SwarnaRoopa

482. **Paradox is a rootstock for ________.**

a) Almond
b) Pecannut
c) Walnut
d) Apricot

483. **Neplus Ultra is a popular variety of __________.**

a) Pecannut b) Apricot
c) Apple d) Almond

484. **Non Pareil is the variety of __________.**

a) Pecannut b) Walnut
c) Almond d) Pistachionut

485. **For flower bud initiation, strawberry requires certain minimum period of __________.**

a) Light b) Darkness
c) Rainfall d) Moisture

486. **Day neutral variety of Strawberry is __________.**

a) SengaSengana b) Tioga
c) Tribute Tristar d) Arking

487. **Albinism is a disorder found in __________.**

a) Guava b) Strawberry
c) Rambutan d) Apple

488. **Malta-blood red is a cultivar of**

a) Sweet Orange b) Mandarin
c) Acid lime d) None of the above

489. **Fire blight of pear is caused by __________**

a) *E.amylovora* b) *E.carotovora*
c) *P.infestance* d) *P.syringae*

490. **Colouring pigment of plum is __________**

a) Carotenoides b) Anthocyanins
c) Betalains d) None of the above

491. **Internal browning of apple is due to __________**

a) injury b) Mg deficiency
c) Both (a) and (b) d) None of the above

492. **Which of the following is non-climacteric fruit ?**

a) Apple b) Pear
c) Pineapple d) Avocado

493. **Richest source of Vitamin C is ________**

a) Sour Cherry b) Barbados Cherry
c) Sweet Cherry d) None of the above

494. **Harvesting period of Loquat is _________.**

a) January-February b) May-July
c) March-May d) February-March

495. **Storage temperature of pomegranate is ____.**

a) 2°C b) 7°C
c) 5°C d) 12°C

496. **Highest productivity of pineapple is in ___________**

a) West Bengal b) Tamil Nadu
c) Kerala d) Himachal Pradesh

497. **Prunacin is an organic compound present in which of the following fruits?**

a) Quince b) Apple
c) Plum d) Peach

498. **African pride is a Annona variety is developed by cross between ________ and ___________**

a) Sugar apple and *Annonareticulata*
b) Sugar apple and *A.glabra*
c) Sugar apple and *A.cherimaya*
d) Sugar apple and *A.squamosa*

499. **Pollinizer for Red delicious Apple is _________**

a) Jonathan b) Ambri
c) Golden Delicious d) Early Shanburry

500. **Type of parthenocarpy in Litchi is ________**

a) Stimulative b) Vegetative
c) Stenospermocarpy d) None of these

501. **Harvesting period of cherry is**

a) April b) July
c) October d) December

502. **Low chilling plum variety is**

a) Santa Rosa b) Titron
c) Beauty d) Methley

503. **Wolliness is a physiological disorder of**

a) Peach b) Plum
c) Pear d) Cherry

504. **Thames Pride introduced from Japan is a variety of**

a) Loquat b) Kiwifruit
c) Avocado d) Mangosteen

505. **Which of the following organic acid is predominantly present in the Carambola?**

a) Oxalic acid b) Citric acid
c) Tartaric acid d) Malic acid

506. **Calyx end rot in grape is due to**

a) Ca deficiency b) Mg deficiency
c) S deficiency d) None of these

507. **Macadamia nut belongs to the family**

a) Ebenaceae b) Actinidiaceae
c) Oxilidaceae d) Proteaceae

508. **Protandry type of dichogamy obtained in**

a) *A.squamosa* b) Fig
c) Pomegranate d) Passion Fruit

509. **Gynodioecious cross pollinated fruit crop is**

a) Grape
b) Fig
c) Banana
d) Kokum

510. **In temperate fruit crops, the "root cutting" are prepared during the month of**

a) January
b) February
c) October
d) December

511. **Abakka is a variety of __________**

a) Guava
b) Grape
c) Pineapple
d) Carambola

512. **Which variety of apple is resistant to all fungal diseases?**

a) Liberty
b) Freedom
c) Ambri
d) Ambred

513. **Gamboge is a physiological disorder of**

a) Durian
b) Mangosteen
c) Pear
d) None of these

514. **Interveinalchlorosis of apple is due to the deficiency of**

a) Bo
b) Mo
c) Mg
d) Cu

515. **Placentia is a variety of**

a) Pecanut
b) Walnut
c) Apple
d) Almond

516. **Heterodichogamy is found in**

a) Walnut
b) Pecanut
c) Almond
d) Chestnut

517. **Which temperate fruit crop require highest chilling requirement?**

a) Apple
b) Pear
c) Almond
d) Cherry

518. **Flying dragon is the most dwarfing rootstock for**
   a) Citrus b) Guava
   c) Apple d) Plum

519. **Which one of the following temperate fruit is earlier in flowering?**
   a) Apple b) Cherry
   c) Peach d) Pear

520. **Self sterile variety of pear is __________**
   a) China pear b) Patharnakh
   c) Leconte d) Magness

521. **Type of self incompatability in loquat**
   a) Sporophytic b) Gametophytic
   c) Both d) None

522. **Origin of custard apple is**
   a) South-east Asia b) Tropical America
   c) China d) India

523. **Highly shade tolerant fruit crop**
   a) Kiwi fruit b) Rambutan
   c) Carambola d) Durian

524. **Dwarf cultivar of peach is**
   a) Floradasum b) Early Grande
   c) Redhaven d) J.H.Hale

525. **Nugget is the dwarf cultivar of**
   a) Apple b) Pear
   c) Peach d) Plum

526. **Mulching is the most important cultural practice in the cultivation of**
   a) Apple b) Pear
   c) Peach d) Strawberry

527. **Training method followed in peach**
   a) Central leader
   b) Open centre
   c) Modified central leader
   d) Multiple system

528. **In phalsa, training is done in which month**
   a) September b) October
   c) November d) December

529. **Which chemical serves as an alternative to caprification in fig**
   a) SADH b) CPA
   c) BOA d) CCC

530. **Which variety of pear has highest TSS**
   a) Punjab Gold b) Punjab Nectar
   c) Redblush d) Jorgonella

531. **Triploid growth curve is found in**
   a) Kiwi fruit b) Pecanut
   c) Tea d) Coffee

532. **Chandlar is the variety of**
   a) Strawberry b) Plum
   c) Peach d) Apple

533. **Fruit cracking in pear is due to the deficiency of**
   a) Ca b) Mg
   c) B d) N

534. **Largest producer of Almond in the world is**
   a) USA b) Japan
   c) China d) Afganistan

535. **Origin of Apple is**
   a) N.America b) S.America
   c) S.E.Asia d) S.W.Asia

536. **Ultra dwarf root stock of apple is**

a) M-9 b) M-27
c) MM-106 d) MM-111

537. **Fig is**

a) Dioecious b) Monoecious
c) Gynodioecious d) Andromonoecious

538. **Kiwi fruit is a**

a) Evergreen vine b) Evergreen shrub
c) Deciduous tree d) Deciduous vine

539. **Most commonly followed propagation method in kiwi fruit**

a) Seeds b) Layering
c) Cuttings d) Stooling

540. **Australian nut is**

a) Walnut b) Pecanut
c) Macadamia nut d) Hazel nut

541. **Arka Ivory is a variety of**

a) Rose b) Gladiolus
c) Jasmine d) Marigold

542. **Petha Kaddu is known as**

a) Pumpkin b) Ridge gourd
c) Smooth gourd d) Ash gourd

Answers

1	a	2	b	3	d	4	a	5	b	6	a	7	a	8	c	9	b	10	a
11	a	12	a	13	b	14	b	15	a	16	c	17	c	18	a	19	c	20	c
21	b	22	a	23	c	24	a	25	c	26	d	27	c	28	a	29	c	30	a
31	b	32	b	33	a	34	c	35	c	36	c	37	a	38	c	39	b	40	c
41	a	42	a	43	d	44	a	45	d	46	d	47	d	48	d	49	c	50	c
51	b	52	a	53	c	54	a	55	d	56	c	57	b	58	a	59	c	60	a
61	d	62	a	63	a	64	d	65	d	66	d	67	d	68	a	69	a	70	a

71 d 72 c 73 c 74 d 75 a 76 b 77 a 78 c 79 a 80 c

81 a 82 b 83 a 84 a 85 a 86 d 87 b 88 b 89 c 90 a

91 a 92 d 93 a 94 c 95 c 96 c 97 a 98 a 99 c 100 b

101 a 102 a 103 a 104 a 105 a 106 b 107 d 108 a 109 c 110 a

111 b 112 a 113 d 114 d 115 a 116 b 117 b 118 b 119 d 120 a

121 b 122 c 123 d 124 b 125 c 126 d 127 c 128 b 129 d 130 b

131 c 132 a 133 b 134 a 135 a 136 b 137 c 138 a 139 b 140 a

141 a 142 b 143 c 144 d 145 a 146 a 147 b 148 c 149 a 150 c

151 c 152 a 153 a 154 b 155 b 156 a 157 a 158 b 159 a 160 c

161 b 162 c 163 c 164 a 165 a 166 c 167 a 168 d 169 b 170 c

171 d 172 a 173 a 174 d 175 b 176 c 177 c 178 b 179 a 180 a

181 a 182 a 183 a 184 a 185 a 186 a 187 a 188 c 189 b 190 a

191 a 192 a 193 c 194 a 195 a 196 a 197 b 198 b 199 b 200 b

201 b 202 b 203 b 204 b 205 a 206 a 207 b 208 d 209 c 210 d

211 a 212 b 213 d 214 b 215 a 216 a 217 d 218 c 219 d 220 b

221 a 222 d 223 c 224 a 225 c 226 d 227 b 228 d 229 b 230 a

231 c 232 d 233 c 234 c 235 a 236 b 237 d 238 c 239 b 240 a

241 a 242 a 243 c 244 b 245 c 246 d 247 b 248 c 249 d 250 c

251 a 252 a 253 a 254 b 255 b 256 c 257 a 258 c 259 b 260 c

261 d 262 c 263 a 264 a 265 d 266 a 267 d 268 a 269 b 270 a

271 a 272 a 273 c 274 b 275 a 276 d 277 a 278 a 279 b 280 c

281 d 282 a 283 a 284 a 285 b 286 a 287 a 288 a 289 b 290 a

291 a 292 b 293 c 294 a 295 d 296 d 297 b 298 d 299 c 300 d

301 a 302 d 303 c 304 a 305 b 306 b 307 a 308 c 309 d 310 c

311 c 312 a 313 b 314 a 315 a 316 a 317 c 318 d 319 a 320 c

321 c 322 c 323 d 324 b 325 b 326 d 327 b 328 c 329 a 330 b

331 d 332 a 333 c 334 a 335 c 336 b 337 a 338 a 339 a 340 a

341 a 342 c 343 c 344 a 345 d 346 a 347 d 348 c 349 b 350 a

351 c 352 a 353 d 354 a 355 a 356 a 357 a 358 c 359 d 360 b

361 b 362 b 363 b 364 b 365 a 366 a 367 a 368 d 369 c 370 c

371 c 372 d 373 b 374 a 375 d 376 d 377 a 378 a 379 a 380 a

381 b 382 b 383 c 384 c 385 c 386 a 387 c 388 b 389 c 390 c

391 b 392 a 393 b 394 a 395 c 396 c 397 d 398 b 399 a 400 c

401 d 402 c 403 a 404 c 405 c 406 a 407 c 408 b 409 b 410 d

411 b 412 c 413 c 414 b 415 b 416 d 417 b 418 c 419 c 420 d

421 b 422 c 423 b 424 b 425 d 426 d 427 a 428 a 429 d 430 b

431 b	432 a	433 c	434 a	435 a	436 b	437 d	438 a	439 a	440 b
441 a	442 a	443 d	444 a	445 b	446 b	447 a	448 b	449 a	450 a
451 c	452 b	453 b	454 c	455 a	456 d	457 b	458 d	459 b	460 d
461 c	462 d	463 b	464 c	465 a	466 a	467 c	468 a	469 c	470 a
471 c	472 b	473 c	474 a	475 c	476 a	477 c	478 c	479 d	480 c
481 d	482 a	483 b	484 c	485 b	486 c	487 b	488 a	489 a	490 a
491 d	492 c	493 b	494 c	495 c	496 a	497 c	498 c	499 c	500 a
501 a	502 b	503 b	504 a	505 a	506 a	507 d	508 d	509 b	510 d
511 c	512 a	513 c	514 b	515 b	516 d	517 a	518 c	519 d	520 b
521 b	522 c	523 c	524 a	525 d	526 b	527 b	528 b	529 b	530 a
531 a	532 a	533 c	534 a	535 d	536 b	537 c	538 d	539 c	540 c
541 a	542 d								

❑❑❑

Chapter 13

Microbiology

1. **Lichen is a symbiotic association between fungi and**
   a) Algae b) Virus
   c) Protozoa d) Plants

2. **Frankia belongs to which of the following groups of microorganisms**
   a) Cyanobacteria b) Green algae
   c) Methanobacterium d) Actinomycetes

3. **Which of the following microorganisms show associative symbiosis?**
   a) Azotobacter b) Azospirillum
   c) Spirulina d) Rhizobium

4. **Streptomycin was discovered by**
   a) Waksman b) Pasteur
   c) Leeuwenhoek d) Winogradsky

5. **Penicillin was discovered by**
   a) Alexander Fleming b) Lederberg
   c) Galileo d) Joseph Lister

6. **Bacteriophage belong to which group of microorganism**
   a) Bacteria b) Virus
   c) Fungi d) Cyanobacteria

7. **The term Rhizosphere was coined by**
   a) Hiltner b) Waksman
   c) Ruinen d) Kary Mullis

8. **The use of condenser in microscope was introduced by**
   a) Abbe b) Pasteur
   c) Adolf Mayer d) Watson and Crick

9. **Blue mold rot in bread is caused by**
   a) Penicillium b) Rhizopus
   c) Mucor d) Salmonella

10. **Spontaneous generation theory was finally disproved by**
   a) Louis Pasteur b) Robert Koch
   c) Paul Ehrlich d) Metchnikoff

11. **Ray fungi is another name of which of the following microorganisms**
   a) Actinomycetes b) Cyanobacteria
   c) Fungus d) Lichen

12. **Phycology deals with the study of**
   a) Algae b) Fungi
   c) Protozoa d) Mycoplasma

13. **Which of the following is of prokaryotic origin**
   a) Blue-Green algae b) Brown algae
   c) Green algae d) Red algae

14. **Blue-green algae belong to a group of microbes called**
   a) Red algae b) Green algae
   c) Brown algae d) Cyanobacteria

15. **Nitrification refers to conversion of ammonia to**
   a) Nitric oxide b) Nitrous oxide
   c) Nitrogen gas d) Nitrate

16. **Mycorrhizae are symbiotic association between Fungi and**
   a) Fungi b) Bacteria
   c) Plant leaves d) Plant roots

17. **Bacterial soft rot of fruit and vegetables is caused by**
    a) Erwiniacarotovora b) Salmonella
    c) Proteus vulgaris d) Leuconostoc

18. **The chemical 'sodium propionate' is generally used as an inhibitor of**
    a) Bacteria b) Mold
    c) Virus d) Algae

19. **Which of the following is an anaerobic nitrogen fixing bacteria?**
    a) Azotobacter b) Salmonella
    c) Derxia d) Clostridium

20. **Which of the following is a symbiotic nitrogen fixing microorganism?**
    a) Frankia b) Klebsiella
    c) Desulphovibrio d) Clostridium

21. **Which of the following is not a entomopathogenic fungus?**
    a) Beauveria b) Verticillium
    c) Hirsutella d) Penicillium

22. **Biogas contains highest amount of which of the following gases**
    a) Methane b) Oxygen
    c) Carbon dioxide d) Hydrogen

23. **Which of the following microorganisms causes Typhus fever**
    a) Rickettsia b) Spiroplasma
    c) Mycoplasma d) Protozoa

24. **Which of the following is an unicellular fungus**
    a) Saccharomyces b) Trichoderma
    c) Glomus d) Gigaspora

25. **Which of the following genetic recombination involves uptake of naked DNA**
    a) Transformation b) Transduction
    c) Conjugation d) Transfection

26. **Which of the following organism convert ammonia to nitrite**
    a) Nitrosomonas b) Nitrobacter
    c) Azospirillum d) Micrococcus

27. **Which of the following organism convert nitrite to nitrate**
    a) Nitrobacter b) Pseudomonas
    c) Clostridium d) Bacillus

28. **Which of the following microorganisms brings about denitrification**
    a) Thiobacillus b) Bacillus
    c) Aspergillus d) Clostridium

29. **Bollworm resistant cotton varieties contains gene from which microorganism**
    a) Bacteria b) Fungi
    c) Protozoa d) Virus

30. **Which of the following is called as extra-chromosomal genetic material**
    a) Plasmid b) Gene
    c) Genotype d) Phenotype

31. **The term plasmid was introduced by**
    a) Lederberg b) Tatum
    c) Beadle d) Kornberg

32. **The genetic material of Viroids consist of**
    a) DNA b) RNA
    c) Proteins d) Both DNA and RNA

33. **Coconut cadang cadang is caused by**
    a) Viroids b) Prions
    c) Protozoa d) Nematodes

34. **Most of the antibiotics are produced by**
    a) Actinomycetes b) Cyanobacteria
    c) Methanobacterium d) Plasmodium

35. **Which of the following is responsible for earthy smell of soil just after rainfall?**
    a) Actinomycetes b) Virus
    c) Bacillus licheniformis d) Bacillus subtilis

36. **Capsid is a part of which of the following microorganism**
    a) Bacteria b) Virus
    c) Algae d) Fungi

37. **Azolla shows symbiotic association with**
    a) Anabaena b) Bacillus
    c) Nostoc d) Calothrix

38. **Heterocysts are important in**
    a) N-fixation b) K-fixation
    c) S-fixation d) P-fixation

39. **Heterocysts are present in which of the following microorganisms**
    a) Cyanobacteria b) Fungus
    c) Actinomycetes d) Streptomyces

40. **Mycorrhiza are important in uptake of**
    a) Phosphorous b) Sulfur
    c) Calcium d) Nitrogen

41. **Siderophores are useful in chelating of which of the following minerals**
    a) Iron b) Magnesium
    c) Phosphorous d) Sulfur

42. **Which of the following is absent in bacteria**
    a) Cell wall b) Cell membrane
    c) Nuclear membrane d) Chromosome

43. **Bacteria do not reproduce by**
    a) Binary fission b) Fragmentation
    c) Asexual reproduction d) Sexual reproduction

44. **TMV (Tobacco Mosaic Virus) contains ……… as genetic material**
   a) DNA b) RNA
   c) Both DNA and RNA d) Proteins

45. **Syphilis is caused by**
   a) Treponema pallidum b) Pseudomonas fluorescence
   c) Bacillus subtilis d) Aspergillus flavus

46. **Who discovered the causative agent of malaria**
   a) Laveran b) Robert Koch
   c) Klebs and Loeffler d) Nicolair

47. **The phenomenon of Ice nucleation involves**
   a) Virus b) Bacteria
   c) Fungus d) Protozoa

48. **N-acetyl muramic acid is present in**
   a) Cell wall of bacteria b) Cell membrane of bacteria
   c) Cell wall of fungus d) Cell membranc of fungus

49. **Which of the following is an obligate intracellular parasite**
   a)Aspergillus niger b) Influenza virus
   c) Nostocmuscorum d) Saccharomyces spp.

50. **Scrapie disease of ship is caused by**
   a) Viroids b) Prions
   c) Mycoplasma d) Spiroplasma

Answers

1 a	2 d	3 b	4 a	5 a	6 b	7 a	8 a	9 a	10 a
11 a	12 a	13 a	14 d	15 d	16 d	17 a	18 b	19 d	20 a
21 d	22 a	23 a	24 a	25 a	26 a	27 a	28 a	29 a	30 a
31 a	32 b	33 a	34 a	35 a	36 b	37 a	38 a	39 a	40 a
41 a	42 c	43 d	44 b	45 a	46 a	47 b	48 c	49 b	50 b

❑❑❑

Chapter 14

Plant Pathology

1. **The sub division ends with**
   a) Mycetes b) Ales
   c) Mycetidae d) Mycotina

2. **Acervuli formed in order**
   a) Moniliales b) Mycelia sterile
   c) Melanconiales d) Sphaeropsidales

3. **Fungal pathogen cause vascular wilt is survive in soil in the form of**
   a) Oospore b) Chlamydospore
   c) Sclerotia d) Conidia

4. **Mottle leaf of citrus is due to deficiency of**
   a) Zinc b) Iron
   c) Boron d) Calcium

5. **Ascus normally contains definite number of ascospores which is**
   a) Four b) Six
   c) Eight d) Two

6. **Loose smut of wheat can be controlled by seed treatment with**
   a) Vitavax b) Agallol
   c) Thiram d) Captan

7. **Emisan is a**
   a) Copper fungicide b) Organomercurial
   c) Sulphur fungicide d) None of above

8. **Indian Phytopathological Society was started by**

   a) B.B.Mundkur b) K.C.Mehta

   c) J.F.dastur d) S.L.Ajrekar

9. **The abnormal increase in the size of plant organ is known as**

   a) Hyperplasia b) Hypertrophy

   c) Both A & B d) Atrophy

10. **The term necrosis indicates**

    a) Death of cell b) Curling

    c) Blightening d) Atrophy

11. **Hyphal structures sent in to the host cell by different fungi to absorb the nutrients are known as**

    a) Appressoria b) Rhizomorph

    c) Haustoria d) Sclerotia

12. **Systemic infection of Albugo candida in crucifers results in to**

    a) Rotting b) Gummosis

    c) Blight d) Hypertrophy

13. **Fungicides which are absorbed in to the system of the plant and move to the remote site of infection are known as**

    a) Systematic b) Fungistatic

    c) Systemic d) Dressers

14. **Which pathogen caused heavy losses to wine industry in France due to its epidemics in 1875**

    a) Phytophthora infestans b) Helminthosporiumoryzae

    c) Plasmopara viticola d) Uncinula necator

15. **Sclerotia of ergot can be removed from the seed by floating them on**

    a) Salt solution b) Glycerol

    c) Mustard oil d) Kerosine

16. **Mycoplasma like organism as plant pathogen for the first time in 1967 by**
    a) Bawden b) Doi et al.
    c) Smith d) Ivanowski

17. **The discovery of _______ mosaic by Mayer in 1886 was the beginning of studies on viruses as incitants of diseases in plants.**
    a) Tomato b) Mungbean
    c) Soybean d) Tobacco

18. **Dr. K. C. Mehta made an outstanding contribution to Plant Pathology by discovering the disease cycle in India**
    a) Cereal rust b) Smuts
    c) Wilts d) Blights

19. **Paddy blast is effectively controlled by spraying of**
    a) Edifenphos b) Chlorothalonil
    c) Mancozeb d) Metalaxyl

20. **Solar energy treatment of wheat seed is recommended for the control of**
    a) Karnal bunt b) Loose smut
    c) Hill bunt d) Earcockle

21. **Generally powdery mildew fungi produced sexual fruiting body is known as**
    a) Apothecium b) Perithecium
    c) Clestothecium d) Pseudothecium

22. **Which of the following rust is the earliest to appear on wheat in India**
    a) Black rust b) Yellow rust
    c) Brown rust d) None of above

23. **Coenocytic mycelium is observed in the**
    a) Higher fungi b) Lower fungi
    c) Imperfect fungi d) None of above

24. **A cabbage disease which is known to be more severe in acidic soils is**
    a) Club root b) yellows
    c) Black rot d) Damping off

25. **Viruses which can killed the bacteria are known as**
    a) Virion b) Antibodies
    c) Bacteriophase d) Viriods

26. **Zoospores of Oomycetes fungi are**
    a) Biflagellate b) Uniflagellate
    c) Tinsel type d) Whiplash type

27. **Resting spore of rust fungi is**
    a) Uredospore b) Teliospore
    c) Basidiospore d) Ascospore

28. **The first scientist who in 1929 studied fungi and saw their spores**
    a) Anton de Bary b) Leeuwenhoek
    c) P.A.Micheli d) Prevost

29. **Downy mildew disease can be effectively managed by spraying of**
    a) Metalaxil MZ b) Thiophanate methyl
    c) Mancozeb d) Carbendazim

30. **Alternate host of pearl millet rust is**
    a) Barberry b) Brinjal
    c) Chaenopodium d) Glycine max

31. **Yellow vein mosaic of okra is transmitted through**
    a) Aphid b) Thrips
    c) Mites d) White fly

32. **Root rot of castor is favoured by**
    a) High soil moisture b) Low soil moisture
    c) High humidity d) Low temperature

33. **A five kingdom system of classification of living beings was suggested by**

a) Linneaus b) Persoon
c) Whittakar d) Haeckel

34. **Bud necrosis of groundnut is caused by**

a) Phytoplasma b) Viroid
c) Spiroplasma d) Virus

35. **Indian Phytopathological society was started by**

a) B.B. mundkur b) K.C. mehta
c) J.F. dastur d) G.S. kulkarni

36. **Viriods spread from cell to cell by**

a) Movement protein b) Plasmodesmata
c) Helper virus d) Cell division

37. **Yellow vein mosaic of okra is caused by**

a) Gemini virus b) Potyvirus
c) Potexvirus d) Comovirus

38. **Express of resistance in plant disease reaction from susceptible appropriate treatment is called**

a) Functional resistance b) Induced resistance
c) Structural resistance d) Synthesis resistance

39. **Plant quarantine method is useful only when**

a) The pathogen is seed transmitted
b) The pathogen is wind borne
c) The pathogen is soil inhabitant
d) The pathogen is vector transmissible

40. **Green ear disease of bajara is also known as**

a) Bunchy top b) Witches broom
c) Downy mildew d) Phyllody

41. **Mango malformation is common in**
    a) North-East India b) North-West India
    c) South-East India d) South-West India

42. **Which one is a host specific toxin?**
    a) Tab toxin b) Ten toxin
    c) Victorin d) Phaseotoxin

43. **Association of fungi with roots of higher plants is known as**
    a) Lichen b) Mycorrhiza
    c) Ascolichen d) Basidiolichane

44. **Zoospores of Oomycetes fungi are**
    a) Biflagellate
    b) Uniflagellate (Tinsel type)
    c) Uniflagellate (whiplash type)
    d) Non motile

45. **The pathogenic organisms included under family Erysiphaceae cause**
    a) Downy mildew b) Powdery mildew
    c) white rust d) ergot

46. **Ascus normally contains definite number of ascospores which is**
    a) Four b) Six
    c) Eight d) Two

47. **When disease is more or less constantly present from year to year in a moderate to severe form, in a particular country or part of the earth then it is known as**
    a) Epidemic b) Sporadic
    c) Pandemic d) Endemic

0

48. **First plant disease caused by parasitic nematode was reported by**
    a) N.A.Cobb b) T.Needham
    c) Prevost d) F.C.Bawden

49. **Pyrimidines fungicides are used against**

a) Rusts b) Powdery mildews
c) Downy mildews d) Smuts

50. **Which of the following fungicide was discovered accidently by Millardet**

a) Vitavax b) Captan
c) Cerasan d) Bordeaux mixture

51. **Karathane is a**

a) Fungicide b) Insecticide
c) Nematicide d) Bactericide

52. **In bacterial brown rot and wilt of potato, the amount of _________ produced by pathogens is proportional to severity of symptoms.**

a) Gibberellin b) Lipids
c) Polysaccharides d) Proteins

53. **Potato leaf roll virus belongs to**

a) Potexvirus group b) Tobravirus group
c) Leuteovirus group d) Potyvirus group

54. **Yellow vein mosaic of okra is transmitted through**

a) Thrips b) Aphids
c) Mites d) Whitefly

55. **Fusarium wilt of banana is also known as**

a) Panama disease b) Moko disease
c) Bunchy top d) Sigatoka disease

56. **Which of the following rust is the earliest to appear on wheat in India**

a) Black rust b) Brown rust
c) Yellow rust d) White rust

57. **Which of the following is an alternate host required by Pucciniagraministritici to complete its life cycle**
    a) Berberis b) Brinjal
    c) Horeum vulgare a) Glycine max

58. **The book entitled "Plant Diseases" has been written by**
    a) Y.L.Nene b) R.S.Singh
    c) Rangaswami d) V.N.Pathak

59. **The most common stage of the fungus causing stem rot of paddy is**
    a) Sclerotia b) Pycnidia
    c) Acervulus d) Myceloid

60. **The most destructive phase of the bacterial blight of rice is known as**
    a) Necrosis b) Kresek
    c) Blight d) Die back

61. **Which one is a polymorphic fungus**
    a) Puccinia graminis tritici
    b) Albugo candida
    c) Alternaria solani
    d) Erysiphe polygoni

62. **Local and systemic infection found in**
    a) Coffee rust b) White rust in crucifers
    c) Late blight of potato d) Early blight of tomato

63. **Basidiospore of black rust are germinated on**
    a) Mustard b) wheat
    c) Barbery d) Rice

64. **Fungi are**
    a) Autotrophic b) Heterotrophic
    c) Phototrophic d) None of above

65. **Which one is a meiospore**
  a) Ascospore b) Sporangia
  c) Conidia d) Chlamydospore

66. **The highest percentage of water (>70%) present in conidia of**
  a) Rusts b) Smuts
  c) Powdery mildews d) Downy mildews

67. **The well proven antifungal antibiotic is**
  a) Penicillin b) Streptomycin
  c) Amphotercin d) Tetramycin

68. **Hemileiavastratrix is a**
  a) Microcyclic b) Macrocyclic
  c) Demicyclic d) Nuctar cycle

69. **Mycoparasitism is the phenomena between**
  a) Fungi and Bacteria b) Fungi and Fungi
  c) Bacteria and Virus d) Fungi and Virus

70. **The term 'Phtophthora' means**
  a) Plant destroyer b) Root rot
  c) Phytonlike d) Potato rot

71. **Clamp connection is found in**
  a) Albugo b) Ustilago
  c) Erysiphe d) Phytophthora

72. **Which one pathogenic fungal genus is polyphagus?**
  a) Alternaria b) Pyricularia
  c) Curvularia d) Venturia

73. **Which one is a soft rot causing fungus?**
  a) Chaetomium b) Coriolus
  c) Xylaria d) Mucor

74. **Rust teliospore on germination produce**

a) Two basidiospore b) Three basidiospore

c) Four basidiospore d) None of the above

75. **Death of tissue in virus -infected plant described as**

a) Mottling b) Mosaic

c) Necrosis d) Curling

76. **Khaira disease of rice is controlled by spraying :**

a) Copper sulphate b) Manganese sulphate

c) Borax d) Zinc sulphate

77. **The disease responsible for the great Bengal famine in 1942-43 is :**

a) Rice blast b) Brown spot of rice

c) Wheat rust d) Take all disease of wheat

78. **'Mosaic' is the specific symptoms produced only by :**

a) Virus b) Phytoplsama

c) Viroid d) Fastidious vascular bacteria

79. **Sooty or charcoal like powdery mass usually appearing on floral organs particularly the ovary is :**

a) Smut b) Bunt

c) Powderymildew d) Sooty mould

80. **Only one disease cycle occurring per season in :**

a) Monocyclic diseases b) Polycyclic diseases

c) Polyetic diseases d) Multiple cycle diseases

81. **The alternate host for pearl millet rust is :**

a) Brinjal b) Mustard

c) Tomato d) Okra

82. **The fungal disease related to the discovery of gibberellins is :**

a) Rice blast b) Foolish seedling disease of rice

c) Rice false smut d) Rice white tip nematode

83. **The direct infection and loss of floral organs as well as seed occurs in :**

a) Karnal bunt of wheat b) Smut of pearl millet
c) Ergot of pearl millet d) All above

84. **Which of the following rust is not caused by fungi ?**

a) Black rust b) Yellow rust
c) Brown rust d) Red rust

85. **Black, yellow and brown rust are generally caused by :**

a) Puccinia spp. b) Uromyces spp.
c) Melampsora spp. d) Phragmidium spp.

86. **A group of diseases caused by basidiomycetes are :**

a) Mildew b) Smut
c) Rot d) Ergot

87. **The typical example of monocyclic diseases is :**

a) Loose smut of wheat b) Rice blast
c) Sesame phyllody d) White rust of mustard

88. **Bacterial rot of wheat ears is also known as :**

a) Spike blight b) Yellow slime disease
c) Tundu d) All above

89. **Brown rust of wheat is caused by :**

a) Pucciniastriiformis b) Pucciniagraministritici
c) Pucciniarecondita d) Ustilagotritici

90. **Loose smut of wheat is a :**

a) Soil borne disease
b) Internally seed borne disease
c) Externally seed borne disease
d) Air borne disease

91. **Grassy shoot disease is caused by :**
   a) Bacteria b) Fungi
   c) Phytoplasma d) Spiroplasma

92. **The pith of the red rot affected cane emits :**
   a) Alcoholic smell b) Fried-egg like smell
   c) Rotten fish like smell d) No smell

93. **Ear showing honey dew symptoms is characteristic feature of :**
   a) Green ear of bajra b) Grain smut of bajra
   c) Ergot of bajra d) All of the above

94. **Which of the following smut of sorghum is not a seed-borne disease?**
   a) Grain smut b) Loose smut
   c) Long smut d) Head smut

95. **Phyllody is caused by :**
   a) Bacteria c) Virus
   b) Fungi d) Phytoplasma

96. **The mode of infection in ergot of bajra is :**
   a) Shoot infection
   b) Seedling infection
   c) Local infection of blossom
   d) Systemic infection

97. **Tikka disease of groundnut is caused by :**
   a) Fusarium c) Colletotrichum
   b) Verticillium d) Cercospora

98. **Which of the following is a pseudo rust ?**
   a) Black rust of wheat b) Brown rust of wheat
   c) Yellow rust of wheat d) White rust of crucifers

99. **The whip smut of sugarcane is caused by :**
   a) Ustilago maydis b) Ustilagoscitaminea
   c) Ustilagohordei d) None of abobe

100. **Which of the following disease of pigeonpea is caused by virus ?**
   a) Sterility mosaic b) Wilt
   c) Stem canker d) Leaf spot

101. **Karnal bunt was first reported by :**
   a) Mitra b) Dowson
   c) Desai d) Butler

102. **Alternate host of black stem rust of wheat in India is :**
   a) Mahonia sp. b) Thalictrum sp.
   c) Berberries sp. d) None of above

103. **Bud-necrosis of groundnut is transmitted by :**
   a) Thrips b) Whitefly
   c) Aphid d) All above

104. **The grain smut of sorghum is also known as :**
   a) Covered smut b) Kernel smut
   c) Short smut d) All above

105. **The wilt of cotton is :**
   a) Seed borne b) Soil borne
   c) Both a & b d) Air borne

106. **Smut disease infect the plant at :**
   a) Seedling stage b) Embryo stage
   c) Tillering stage d) None of above

107. **Solar heat treatment is used to control :**
   a) Covered smut of barley b) Loose smut of wheat
   c) Smut of bajra d) Smut of sorghum

108. **The first recognized virus disease of plants in world is :**
   a) Tobacco mosaic b) Cucumber mosaic
   c) Yellow vein mosaic d) None of above

109. **The annual reoccurrence of rust of wheat (black rust) in the plains is through :**
   a) Uredospores c) Aeciospores
   b) Teleutospores d) Basidiospores

110. **The incidence of which disease increases with increased supply of nitrogen?**
   a) Rice blast b) Cotton wilt
   c) Potato tuber rot d) All above

111. **Which stage of the wheat rust fungus is considered as the perfect stage?**
   a) Uredial stage b) Telial stage
   c) Pycnial stage d) None of above

112. **Sugarcane crop appears dry and on splitting the cane white and red bunds are seen with foul smell. The disease is :**
   a) Red rot b) Whip smut
   c) Wilt d) Grassy shoot

113. **The most serious smut among the smuts affecting sorghum is :**
   a) Grain smut c) Long smut
   b) Loose smut d) Head smut

114. **Frog-eye leaf spot disease is present in :**
   a) Tobacco c) Cotton
   b) Potato d) None of above

115. **In which rust, pustules are arranged in end to end manner and forming stripes ?**
   a) Yellow rust b) Orange rust
   c) Black rust d) White rust

116. **The symptoms of red rot of sugarcane mostly found on :**
   a) Stem b) Root
   c) Leaf d) Arrow

117. **Leaf curl of tobacco is caused by :**
   a) Nicotiana virus -1 b) Nicotiana virus -10
   c) Phytoplasma d) Viroid

118. **Puccinia causes rust on :**
   a) Cereals b) Flowers
   c) Vegetables d) All above

119. **Root parasite of Bajra is :**
   a) Cuscuta b) Striga
   c) Orobanche d) Loranthus

120. **Pegion pea sterility mosaic disease is transmitted by**
   a) Aphid b) Eryophide mite
   c) Whitefly d) Jassid

121. **Phyllody of sesamum is transmitted by**
   a) Adult thrips b) Aphid
   c) Adult jassid d) Mite

122. **Khaira disease of rice is controlled by spraying:**
   a) Copper sulphate b) Manganese sulphate
   c) Borax d) Zinc sulphate

123. **The disease responsible for the great Bengal famine in 1942-43 is :**
   a) Rice blast b) Brown spot of rice
   c) Wheat rust d) Take all disease of wheat

124. **The whip smut of sugarcane is caused by:**
   a) *Ustilago maydis* b) *Ustilago scitaminea*
   c) *Ustilago maydis* d) *Ustilago scitaminea*

125. **Partial root parasite of sorghum is:**

a) Cuscuta b) Striga
c) Orobanche d) Loranthus

126. **Total root parasite of tobacco is**

a) Cuscuta b) Striga
c) Orobanche d) Loranthus

127. **Mungbean yellow mosaic virus is secondary transmitted by**

a) Whitefly b) Thrips
c) Jassid d) Aphid

128. **Root knot disease of tobacco is effective control by**

a) Carbendazim b) Carbofuran
c) Thirum d) Mancozeb

129. **Stem rot of groundnut pathogen survives in the form of**

a) Sclerotia b) Chlamydospores
c) Oospores d) Conidia

130. **Which insect is transmitted bud necrosis of groundnut?**

a) Aphid b) Thrips
c) Jassid d) Mite

131. **Which type of soil is most favored for root knot nematode ?**

a) Sandy soil b) Clay soil
c) Heavy black d) Calcareous soil

132. **Rhizome rot pathogen of turmeric is transmitted by**

a) Rhizome b) Leaf
c) Air d) Non of them

133. **White rust of mustard is caused by**

a) *Puccinia graminis* b) *Puccinia arachidis*
c) *Puccinia recondita* d) *Albugo candida*

Answers

1 d	2 c	3 b	4 a	5 c	6 a	7 b	8 a	9 c	10 a
11 c	12 d	13 c	14 c	15 a	16 b	17 d	18 a	19 a	20 b
21 c	22 c	23 b	24 a	25 c	26 a	27 b	28 c	29 a	30 b
31 d	32 b	33 c	34 d	35 a	36 b	37 a	38 b	39 a	40 c
41 b	42 c	43 b	44 a	45 b	46 c	47 d	48 b	49 b	50 d
51 a	52 c	53 b	54 d	55 a	56 b	57 a	58 b	59 a	60 b
61 a	62 b	63 c	64 b	65 a	66 c	67 c	68 a	69 b	70 a
71 b	72 a	73 a	74 c	75 c	76 d	77 b	78 a	79 a	80 a
81 a	82 b	83 b	84 d	85 a	86 b	87 a	88 d	89 c	90 b
91 c	92 a	93 c	94 b	95 d	96 c	97 d	98 d	99 b	100 a
101 a	102 c	103 a	104 d	105 c	106 b	107 b	108 a	109 a	110 d
111 b	112 a	113 a	114 a	115 a	116 a	117 a	118 a	119 b	120 b
121 c	122 d	123 b	124 d	125 b	126 c	127 a	128 b	129 a	130 b
131 a	132 a	133 d							

□□□

Chapter 15

Soil Science

1. **It is the relative purity or strength of the spectral colour.**

   a) Hue b) Light
   c) Chroma d) Value

2. **"The limiting velocity of a particle falling in fluid medium will be proportional to the square of its diameter." This low is given by:**

   a) Beer b) Jackson
   c) Stoke d) Lambert

3. **This enzyme is involved in the BNF.**

   a) Rhizobium b) Azotobacter
   c) Nitrogenase d) Nitrosomonas

4. **"Enrichment of water bodies by nutrients and consequent bloom of algal and other aquatic plants" This is the process known as:**

   a) Illuviation b) Elluviation
   c) Eutrophication d) Neutralization

5. **A branch of soil science which deals with the soil from the crop production point of view**

   a) Pedology b) Soil geology
   c) Ecology d) Edaphology

6. **In montmorillonite, most of the negative charges are developed by:**

   a) Dissociation of OH group
   b) Ionic substitution
   c) Ionic structure
   d) Humus

7. **Particle density is also known as:**
   a) True density b) Bulk density
   c) Aopparent density d) Specific gravity

8. **Total porosity is more in**
   a) Sandy soil b) Clayey soil
   c) Silty soil d) Loamy sand

9. **The dominant potential in saline soil is**
   a) Metric b) Osmotic
   c) Gravitational d) pressure

10. **The phenomenon by which hydrogen atoms act as links between water molecules is called as**
   a) Polarity b) Hydrogen bonding
   c) Surface tension d) Attraction

11. **The attraction of solid surface for water molecules is called as**
   a) Adhesion b) Cohesion
   c) Flocculation d) Plasticity

12. **The best agricultural soil structure is**
   a) Crumb b) Platy
   c) Columnar d) Bloky

13. **If igneous rock contains more than 65 per cent silica it is called as**
   a) Acidic b) Intermediate
   c) Basic d) Ultrabasic

14. **The soil having more than 30% organic matter is placed in**
   a) Oxisols b) Aridisols
   c) Histosols d) Alfisols

15. **The example of 2 : 2 type silicate clay mineral is**
   a) Chlorite b) Kaolinite
   c) Montmorillonite d) Vermiculite

16. **The process of desilication is known as**
    a) Laterization b) Podzolization
    c) Salinization d) Humification

17. **The main centre of NBSS & LUP is located at**
    a) Palanpur b) Nagpur
    c) Kanpur d) Jodhpur

18. **The element N is not found in**
    a) Atmosphere b) Hydrosphere
    c) Lithosphere d) Planets

19. **The metamorphic rock formed due to action of high pressure is known as**
    a) Dynamo metamorphic b) Thermo metamorphic
    c) Hydro metamorphic d) None of these

20. **The progressive accumulation of Na^{+} ion on exchange complex is called as**
    a) Salinization b) Calcification
    c) Alkalization d) Gleization

21. **The soils formed in high rainfall area are**
    a) Acidic b) Basic
    c) Alkaline d) Saline

22. **Horizon B_2 is also known as**
    a) Illuvial horizon b) Eluvial horizon
    c) Solum d) Transition horizon

23. **Hematite is converted to limonite by the process of**
    a) Hydration b) Hydrolysis
    c) Oxidation d) Carbonation

24. **Name of phosphorus supplying mineral is**
    a) Quartz b) Apatite
    c) Feldspar d) Calcite

25. **The horizon A + B are collectively called as**

a) Solum
b) Regolith
c) Organic
d) Illuvial

26. **The order of desert soil is**

a) Entisoils
b) Aridisols
c) Vertisols
d) Inceptisols

27. **The best agricultural texture is**

a) Loam
b) Sand
c) Silt
d) Clay

28. **The water retained by air dried soils is considered as**

a) Gravitational
b) Hygroscopic
c) Capillary
d) Free

29. **Natural soil aggregates are known as**

a) Peds
b) Clods
c) Floccules
d) None of these

30. **Soil pH is a measure of**

a) Active acidity
b) Reserve acidity
c) Exchange acidity
d) None of these

31. **Infiltration rate is relatively higher in**

a) Sandy soil
b) Clay loam soil
c) Silty soil
d) Clayey soil

32. **Widely accepted fixed value of particle density is**

a) 1.5 g/cc
b) 2.65 g/cc
c) 2.29 g/cc
d) 1.65 g/cc

33. **Solution whose strength or concentration is accurately known is termed as**

a) Normal solution
b) Concentrate solution
c) Standard solution
d) solvent

34. **Study of origin, classification, morphology of soil is known as**
    a) Pedology b) Edaphology
    c) Soil physics d) Physical chemistry

35. **Solution which contains one gram eq. wt. of substance dissolved in one lit. of solution is known as**
    a) Molar solution b) Normal solution
    c) Molal solution d) Standard solution

36. **The size of silt fraction according to USDA system is**
    a) 0.05-0.002 mm b) 0.02-0.002 mm
    c) < 0.02 mm d) 0.2–0.02 mm

37. **The range of usefulness of tensiometer is between**
    a) 10-15 bar b) 1-10 bar
    c) 0.0-0.8 bar d) - 15 bar

38. **An average solar radiation reaching to the earth is**
    a) 80 % b) 70 %
    c) 60% d) 50%

39. **Maximum density of water is at**
    a) 4 °C b) - 4 °C
    c) 0 °C d) 10 °C

40. **The water content on mass basis can be converted into volume basis by multiplying it with**
    a) Particle density b) Surface tension
    c) Bulk density d) Total potential

41. **Which one is considered as organic fertilizer?**
    a) FYM b) Urea
    c) Ammonium sulphate d) DAP

42. **The process by which ions are taken into plant roots**
    a) Oasis b) Absorption
    c) Advection d) Adsorption

43. **An oxidative transformation of NH_4^+ into NO_2^- is mediated by**

a) Nitrobacter
b) Urease
c) Nitrosomonas
d) Nitrogenase

44. **A by product of coastal salt industry- schoenite is a rich source of plant nutrient**

a) N
b) P
c) K
d) Zn

45. **It has higher CEC**

a) FYM
b) Humus
c) Root
d) Clay

46. **The C : N ratio of the soil are fairly constant between**

a) 10 : 1 to 12 : 1
b) 50 : 1 to 60 : 1
c) 1 : 10 to 1 : 12
d) 1 : 10 to 12 : 6

47. **Rock phosphate is applied in**

a) Normal soil
b) Calcareous soil
c) Acid soil
d) Degraded alkali soil

48. **Conversion factor for P to P_2O_5 is**

a) 2.00
b) 2.5
c) 2.29
d) 1.92

49. **According to LCC, which class of land is suitable for wild life**

a) III
b) VIII
c) IV
d) VI

50. **The essentiality of N was established by**

a) E. Gris
b) De Saussure
c) Arnon
d) Lipman

51. **The gram equivalent weight of Na_2CO_3 is**

A) 40
B) 53
C) 106
D) 50

52. The gram equivalent weight of H_2SO_4 is

a) 98 b) 36.5

c) 49 d) 63

53. One angstrom is equal to

a) 10^{-8} m b) 10^{-9} m

c) 10^{-7} m d) 10^{-10} m

54. One gram equivalent NaOH is equal to

a) 1000 mg NaOH b) 40 g NaOH

c) 4 g NaOH d) 10000 mg NaOH

55. One milli equivalent NaOH is equal to

a) 40 mg NaOH b) 4 g NaOH

c) 1000 mg NaOH d) 4000 mg NaOH

56. Sodium chloride is a salt of

a) Strong acid – weak base

b) Weak acid – strong base

c) Strong acid – strong base

d) Weak acid – weak base

57. pH may be defined as

a) Negative log of OH ion concentration

b) Log of H ion concentration

c) Negative log of H ion concentration

d) Log of OH ion concentration

58. Basicity of H_3PO_4 is

a) 1 b) 2

c) 3 d) 4

59. Acidity of $Ca(OH)_2$ is

a) 1 b) 2

c) 3 d) 4

60. **Basicity of oxalic acid ($H_2C_2O_4.2H_2O$) is**

a) 1 b) 2
c) 6 d) 3

61. **If the solution contains [OH^-] = 10^{-8} g mole/l than pH of this solution will be**

a) 6 b) 8
c) 2 d) 4

62. **Parts per million is equal to**

a) g/l b) mg/l
c) g/kg d) mg/ml

63. **Oxidation reaction involves**

a) Loss of electrons b) Gain of electrons
c) Sharing of electrons d) None of these

64. **The example of oxidizing agent is**

a) NaCl b) $KMnO_4$
c) $Ca(OH)_2$ d) NaOH

65. **The example of self indicator is**

a) Phenolphthalein b) Methyl orange
c) $KMnO_4$ d) Methyl red

66. **The gram equivalent weight of $KMnO_4$ is**

a) 158 b) 31.6
c) 63.2 d) 94.8

67. **The glass electrode is used in**

a) pH meter b) E. C. Meter
c) Colorimeter d) Spectrometer

68. **Filter is used as dispersing device in**

a) Spectrometer b) Colorimeter
c) pH meter d) E. C. Meter

69. **Beriberi is the disease caused by the deficiency of vitamin**

a) A
b) B_1
c) B_6
d) B_{12}

70. **Chlorophyll contains**

a) Magnesium
b) Manganese
c) Iron
d) Cobalt

71. **Hemoglobin contains**

a) Nickel
b) Cobalt
c) Iron
d) Magnesium

72. **The method used to determine available phosphorus from soil is**

a) Moh'r method
b) Versenate method
c) Olsen's method
d) Molisch's method

73. **Fehling's B solution contains salt known as**

a) Raswlf salt
b) Benedict salt
c) Rochelle salt
d) Rock salt

74. **Vitamin C is also known as**

a) Acetic acid
b) Formic acid
c) Ascorbic acid
d) Aspartic acid

75. **When diacid mixture is used for digestion of plant sample, the element can not be determined from that extract is**

a) N
b) K
c) P
d) S

76. **Fehling's A solution contains**

a) $FeSO_4.5H_2O$
b) $MgSO_4.3H_2O$
c) $CuSO_4.5H_2O$
d) $ZnSO_4.7H_2O$

77. **Method used for the determination of lime requirement of an acid soil is**

a) Moh'r method
b) Shoemaker's method
c) Olsen's method
d) Kleinal's method

78. **The dye used for titrimetric determination of ascorbic acid is**
   a) Para Nitro phenol
   b) 2,4-dinitrophenol
   c) 2-6 di chloro phenol indophenol
   d) 2, 6 dibromo phenol indophenol

79. **At which wavelength, plant phosphorus can be estimated on Uv- vis spectrophotometer?**
   a) 440 nm b) 660 nm
   c) 470 nm d) 570 nm

80. **Metaphosphoric acid is**
   a) H_3PO_4 b) H_3PO_3
   c) HPO_3 d) $H_4P_2O_7$

81. **The indicator used in determination of Organic carbon from soil is**
   a) Methylene blue b) Diphenyl amine
   c) Methyl orange d) Methyl red

82. **Determination of Fe, Mn, Zn, and Cu from plant is donc by using**
   a) FPM b) SPM
   c) AAS d) EC meter

83. **Which vitamin contains metal ion?**
   a) Vitamin A b) Vitamin C
   c) Vitamin B_{12} d) Vitamin B_1

84. **Soil productivity is basically a:**
   a) Social concept b) Economic concept
   c) Physical concept d) Chemical concept

85. **Gypsum is used for the reclamation of**
   a) Sodic soil b) Heavy soil
   c) Saline soil d) Acidic soil

86. **Decomposition of organic matter results in an increase of**
   a) Salinity b) Alkalinity
   c) Acidity d) Sodicity

87. **kg/ha = ppm x ________**
   a) 2.24 b) 4.48
   c) 1.12 d) 1.24

88. **The organic fertilizer is:**
   a) Sufala b) Urea
   c) DAP d) Muriate of potash

89. **The pH value varies from**
   a) 7.00 to 14.00 b) 0.00 to 7.00
   c) 0.00 to 14.00 d) None of these

90. **The green manure crop is turned into soil at the stage of crop**
   a) Germination b) Harvest
   c) Flowering d) Boot stage

91. **Requirement of phosphorus by the crop is maximum at ______ growth stage**
   a) Initial b) Flowering
   c) Lateral d) Harvest

92. **ppb = ___________**
   a) mg/ml b) g/ml
   c) μg/l d) g/l

93. **The salt used along with $CuSO_4$ and $FeSO_4$ as digestion mixture used for N estimation is**
   a) $MgSO_4$ b) K_2SO_4
   c) Na_2SO_4 d) $ZnSO_4$

94. **Vitamin A deficiency causes**
   a) Scurvy b) Beriberi
   c) Night blindness d) Dizziness

95. **Sugar presents in milk**
   a) Lactose b) Galactose
   c) Maltose d) Fructose

96. **Paddy crop absorbs nitrogen in form of:**
   a) NH_2 b) NH_4
   c) NO_3 d) N_2

97. **Organic nitrogen converted in to NH_4 form is known as**
   a) Nitrification b) Ammonification
   c) Denitrification d) Immobilization

98. **Optical Density =**
   a) 2-logT b) Log -2T
   c) 2log-T d) T-log2

99. **The specific gravity of water**
   a) 1.5 Mg m^{-1} b) 1.0 Mg m^{-1}
   c) 5.1 Mg m^{-1} d) 0.1 Mg m^{-1}

100. **The nutrient required for quality maintenance in potato:**
   a) Zinc b) Potassium
   c) Iron d) Phosphorus

101. **The two non-metallic elements occurring in greatest abundance in the earthcrust are**
   a) Calcium & Magnesium b) Oxygen& Silicon
   c) Aluminum & iron

102. **Science which deals with the description of rocks**
   a) Petrography b) Petrogenesis
   c) Chromatography d) pedology

103. **Water plays an importantrole in the formation of**
   a) Igneous rock b) Sedimentary rock
   c) Metamorphic rock d) None

104. **The time that nature devotes to the formation of soil is termed as**

a) Periodical table b) Pedological time
c) Soil stages d) Temporal

105. **When igneous rocks are become metamorphic rocks due to pressure**

a) Hydrometmorphic b) Thermometamorphic
c) Dynamometamorphic d) None

106. **The parent materials deposited near the base of strong slope by the action of gravity**

a) Lucustrain b) Colluvium
c) Alluvium d) Aeolian

107. **The parent materials of sand size transported by the action of wind is called**

a) Eolian b) Moraine
c) Loess d) Alluvium

108. **The excavation and destructive action of water on rock and minerals is called**

a) Exfoliation b) Denudation
c) Chemical weathering d) Marine

109. **The rocks gets broken in pieces due to temperature is called**

a) Chemical weathering b) Denudation
c) Exfoliation d) Physical weathering

110. **The fourth most abundant element in the earth crust is**

a) Oxygen b) Iron
c) Aluminium d) silica

111. **Marble is metamorphic rock converted from lime stone due to the action of**

a) Temperature b) Pressure
c) Heat d) Water

112. **The individual grains size in rudaceous sedimentary rock are of**

a) Sand b) Clay
c) Boulders and pebbles d) Silt

113. **The clay fraction of the soil has diameter less than**

a) 0.002 mm b) 0.2 mm
c) 0.02 mm d) 0.0002 mm

114. **The time that nature devotes to the formation of soil is termed as**

a) Periodical table b) Pedological time
c) Soil stages d) Temporal

115. **B_1 horizon in the soil profile is also known as**

a) Allivial b) Eluvial
c) Illuvial d) Aeolium

116. **Sedimentary rocks are also known as**

a) Metamorphic rock b) Stratified rock
c) Igneous rock d) Extrusive rocks

117. **Accumulation of Calcium carbonate in the profile is called**

a) Salinization b) Solonization
c) Cacification d) Solodization

118. **An active soil forming factor is**

a) Climate b) Parent material
c) Topography d) Time

119. **An artificially formed soil mass is called**

a) Ped b) Crust
c) Clod d) Land

120. **Deep black soil of Gujarat is also known as**

a) Black soil b) Alluvial soil
c) Regurs soil d) Brown soil

121. **Two water molecules are held by ________ bond.**

a) Hydrogen b) Ionic

c) Covalent d) Negative

122. **The soil having PD 2.50 g/cc and BD 1.25 g/cc will have _____ % porosity.**

a) 45 b) 50

c) 55 d) 60

123. **Total porosity is more in:**

a) Sandy soil b) Loamy soil

c) Clayey soil d) Silty soil

124. **The best agriculturalsoilstructure is:**

a) Spheroidal b) Platy

c) Prism d) Columnar

125. **Chemically most active soil separate is**

a) Sand b) Silt

c) Clay d) Gravel

126. **The common clay mineral found in silt is**

a) Montmorillonite b) Mica

c) Quartz d) Kaolinite

127. **The lightness or darkness of soil coulour is denoted by:**

a) Hue b) Value

c) Chroma d) None of these

128. **Black soil are __________ in organic matter:**

a) Medium b) Poor

c) High d) None of these

129. **At FC moisture is held at:**

a) - 31 water potential bar

b) - 1/3 water potential bar

c) - 15 water potential bar

d) None of these

130. **Organic matter influences water holding capacity of soil in the following manner:**
a) Is not affected at all b) Is decreased
c) Is stabilized d) Is increased

131. **Soil containing high proportion of organic matter have the following colour:**
a) Dark colour b) Yellow colour
c) Red colour d) Light colour

132. **The components of soil are:**
a) Mineralmatter b) Organic matter
c) Water & Air d) All of these

133. **Relative proportion of soil separates in sand, silt and clay is termed as:**
a) Soil texture b) Soil structure
c) Soil porosity d) None of these

134. **The arrangement of individual soil particles in a specific pattern is called:**
a) Soil texture b) Soil separates
c) Soil structure d) None of these

135. **When organic matter is added to the soil:**
a) Bulk density does not change
b) Bulk density is decreased
c) Bulk density is increased
d) None of these

136. **A branch of soil science which deals with the soil from the crop production point of view:**
a) Pedology b) Edaphology
c) Ecology d) Soil geology

137. **An average particle density of soils is:**
a) 1.5 g/cc b) 2.0 g/cc
c) 2.65 g/cc d) None of these

138. **The process of determining the amount of sand, silt and clay is known as:**
   a) Soilanalysis b) Mechanical analysis
   c) Statistical analysis d) Chemical analysis

139. **Particle density is also known as:**
   a) True density b) Bulk density
   c) Apparent density d) Specific gravity

140. **The best agricultural soil structure is:**
   a) Crumb b) Platy
   c) Columnar d) Blocky

141. **Volumetrically soil contains __________ % mineral matter.**
   a) 25 b) 5
   c) 45 d) 40

142. **The soil colours are determined by:**
   a) Munsell chart b) Periodic table
   c) Fertility chart

143. **The reaction of oil/fat with NaOH/KOH as.....**
   a) Hydrogenation b) Saponification
   c) Halogenation d) Hydrolysis

144. **Monosaccharids possess reducing property due to the presence of**
   a) Free aldehyde or keto group
   b) Free alcohol group
   c) Free amino group
   d) Free thiol group

145. **Compounds which in solution rotate the plane of polarized light to the right are called**
   a) Laevo rotatory b) Dextro rotatory
   c) Optically inactive d) All of three

146. **At isoelectric point the proteins are**

a) Acidic b) Neutral
c) Basic d) None of these

147. **Enzymes which exist in multiple forms within single specing of an organism**

a) Isoenzyme b) Coenzyme
c) Apoenzyme d) Holoenzyme

148. **Chlorophyll is present in**

a) Vacuole b) Nucleus
c) Chloroplast d) Golgi body

149. **Man cannot digest cellulose due to the absence of**

a) Hemicellulase b) Protease
c) Lipase d) Cellulase

150. **Number of hydrogen bonds present between adenine and thymine**

a) 1 b) 2
c) 3 d) 4

151 **Socrates was given a cup of**

a) Dhatura alkaloid b) Nicotine
c) Stachydrine d) Hordinine

152. **ATP is produced in the cristae of**

a) Mitochondria and chloroplasts
b) Stroma
c) Nucleus
d) Cell wall

153. **The cation exchange phenomenon was first recognized by ________.**

a) Walkly and Black b) Thompson and Way
c) Arnon and Stout d) William and Tan

154. ________ hydroxide act as cementing agent in binding the soil particles together.

a) Ca & Mg b) Fe & Al

c) Na & K d) Ca & Na

155. _________ reagent is used for the determination of micronutrient cations from the soil.

a) EDTA b) CDTA

c) DTPA d) TEA

156. _____ fertilizers are applied for improvement of quality and strength of fiber crops.

a) Phosphatic b) Nitrogenous

c) Potassic d) sulphur

157. ______ process depends on partial dissociation of water into H and OH ions.

a) Hydrolysis b) Hydration

c) Solution d) Carbonation

158. ________ is the ability of a substance to transfer heat from molecule to molecule.

a) Specific heat b) Heat capacity

c) Thermal conductivity d) Thermal diffusivity

159. _________ is non-symbiotic N-fixer in soil.

a) Rhizobium b) Fungi

c) Azatobactor d) Azospirillium

160. _________ is symbiotic N-fixer in leguminous crop.

a) Fungi b) Azatobactor

c) Rhizobium d) Azospirillium

161. _________ potential is due to presence of solute in the soil.

a) Gravimetric b) Osmotic

c) Matric d) pressure potential

162. __________ **is conversion of soil nitrate into gaseous nitrogen.**
   a) Denitrification b) Nitrification
   c) Ammonification d) Denitrification

163. __________ **is one of the phosphate solubilizer species of micro organisms.**
   a) Azolla b) Pseudomonas
   c) Azotobactor d) Azospirillium

164. __________ **is the study of origin of rocks.**
   a) Pedology b) Petrology
   c) Petrograpgy d) Edapology

165. __________ **cation is dominant in sodic soil.**
   a) Sodium b) Potassium
   c) Calcium d) Magnesium

166. __________ **is passive soil forming factor.**
   a) parent material b) Biosphere
   c) climate d) None of these

167. __________ **is example of silicate clay minerals.**
   a) Kaolinite b) Apatite
   c) Tourmaline d) Illite

168. __________ **is fast decomposing material in soil.**
   a) Sugar b) Lignin
   c) Fats d) Protein

169. __________ **mineral is a source of boron in soils.**
   a) Apatite b) Tourmaline
   c) Biotite d) Illite

170. __________ **mineral is a source of phosphorus in soils.**
   a) Apatite b) Tourmaline
   c) Biotite d) Kaolonite

171. ____________ **fertilizers are applied for improvement of quality and strength of fiber crop.**

a) Phosphoric b) Nitrogenous
c) Potassic d) Sulphur

172. _____________ **cake has highest nitrification rate.**

a) Groundnut b) Castor
c) Neem d) Mahua

173. _____________ **is an example of the mostpersistent herbicides.**

a) Simazine, B) 2-b4-D
c) Glyphosate d) 2-4-5-D

174. _____________ **is example of immobile element in soil.**

a) Nitrogen b) Potassium
c) Phosphorus d) Zinc

175. ______________ **fertilizer is used for the paddy crop.**

a) Nitrate b) Ammonical
c) Ammonical –Nitrate d) Amide

176. ______________ **is one of the phosphate solubilizer species of micro organisms.**

a) Azolla b) Pseudomonas
c) Azotobacter d) Rhizobium

177. _________________ **= (CO_3 + HCO_3) – (Ca + Mg).**

a) ESP b) SAR
c) RSC d) RSBC

178. _____________________ **is a recent soil order.**

a) Inceptisols b) Endisols
c) Alfisols d) Aridisol

179. _____________________ **is also known as saline soil.**

a) White alkali b) Black alkali
c) Usar

180. ____________________ **is beneficial element for the plant.**
   a) Nitrogen b) Silicon
   c) Iron d) Phosphorus

181. ____________**is the process of decomposition of organic matter.**
   a) Humification b) Aminization
   c) Nitrification d) Denitrification

182. ____________**structure provides less porosity in soils.**
   a) Platy b) Spherical
   c) Granular d) Columnar

183. **1 m.e. H^+ = ______ mg Ca^{++}.**
   a) 20 b) 40
   c) 12 d) 25

184. **1 me gypsum per 100 g = ________ mg/100g gypsum.**
   a) 86.0 b) 172
   c) 50 d) 75

185. **1 me Na^+ = ______ mg Na^+.**
   a) 23 b) 12
   c) 46 d) 58

186. **A simple method for the determination of moisture is ________**
   a) Gravimetric b) Titration
   c) Neutron scattering d) Cohesion

187. **Accumulation of ________ in soil profile is more in calcification processes.**
   a) $CaCO_3$ b) $CaSO_4$
   c) $CaCl_2$ d) $Nacl_2$

188. **Ammonification involves transformation of**
   a) Sugar to CO_2
   c) Nitrate to Nitrite

c) Amino acid to ammonia

d) Ammonia to amino acid

189. **As per WHO, the safe limit of nitrate content in ground water is _______ mg/l.**

a) 1 b) 10

c) 100 d) 50

190. **At a point C:N ratio, becomes more or less constant, generally stabilizes at_____.**

a) 10:1 b) 20:2

c) 30:3 d) 50:1

191. **Average nitrogen concentration in plant tissue is________ %.**

a) 1.5 b) 0.5

c) 5.1 d) 50

192. **Azatobacter can fix ________ kg atmospheric nitrogen /ha to the soil.**

a) 200 b) 2

c) 20 d) 2000

193. **Boron content of irrigation water ranged from ___ mostly sensitive to citrus crop.**

a) 3.0 – 10.0 ppm b) 0.3 – 1.0

c) 30.0-100.0 d) 30-300

194. **Burning quality of Tobacco decreased due to**

a) Chloride b) Sulphate

c) Potassium d) Phosphorus

195. **CAN fertilizer is _______ fertilizer**

a) Acidic b) Basic

c) Neutral d) None of these

196. **Carryover fertilizer effects are more pronounced in the case of __________.**

a) Urea b) CAN

c) SSP d) DAP

197. **Chemically the most active soil separate is________.**

a) Sand b) Silt

c) Clay d) Gravel

198. **Clay having ____________ charge so that they attract and adsorb primarily cations.**

a) Positive b) Negative

c) Neutral d) None of these

199. **Compaction ________ bulk density of soil.**

a) Increases b) Decreases

c) None of these d)

200. **Conversion of K to K_2O is ______________.**

a) 2.1 b) 1.2

c) 2.2 d) 1.32

201. **Conversion of P to P_2O_5 is ______________.**

a) 2.10 b) 1.20

c) 2.29 d) 2.24

202. **DDT has ______________ half life years in cultivated soils.**

a) 6.0 b) 3.0

c) 9.0 d) 12

203. **The activities of __________ more in acid soils.**

a) Bacteria b) Actinomycetes

c) Fungi d) Algae

204. **The value of soil EC indicates ________ in soil.**

a) H ion b) Insoluble salts

c) Soluble salts d) Al ion

205. **Deficiency of _____ nutrient shows marginal burning of leaves, irregular fruit development of plant.**

a) N b) P

c) K d) Fe

206. **Deficiency of __________ appears short internodes in plant.**

a) Zinc b) Nitrogen

c) Copper d) Potassium

207. **Di-Ammonium Phosphate fertilizer contains ________ % nitrogen.**

a) 46 b) 21

c) 18 d) 24

208. **Eaton (1950) suggested the concept of ___________.**

a) RSC b) SAR

c) TSS d) RSBC

209. **Essentially criteria of nutrient are given by _______________.**

a) Schoonover b) Arnon

c) Jackson d) Schofield

210. **Exchangeable sodium percentage (ESP) less than __________ is safe for tolerable soil physical conditions.**

a) **15** **b) 20**

c) 30 d) 25

211. **Fertilizer like ___________ destroys the soil aggregates.**

a) Urea b) Sodium nitrate

c) Calcium nitrate d) DAP

212. **Flame-photometer is used for the determination of ___________.**

a) Available P b) Available K

c) Available N d) Sulphur

213. **For multi element analysis, the plant materials should be digested in _________acid alone.**

a) Nitric
b) Sulphuric
c) Perchloric
d) Hydrochloric

214. **Generally, _____ extractant is used to determine available micronutrients from soil.**

a) EDTA
b) CDTA
c) DTPA
d) $CaCl_2$

215. **1 me Na^+ = ______ mg Na^+.**

a) 23
b) 12
c) 46
d) 58

216. **Generally, true density of normal soils is __________ g/cc.**

a) 1.65
b) 2.65
c) 3.65

217. **Growth stimulation of aquatic organisms through nutrient enrichment of lakes and ponds is known as ____________.**

a) Enrichment
b) Amendment
c) Eutrofication

218. **High nitrate content in drinking water causes ________ disease in infants.**

a) Anemia
b) Methamoglobinemia
c) Pneumonia

219. **Illite is ___________ type of silicate clay mineral.**

a) 2 : 1
b) 1 : 1
c) 1 : 2

220. **Immobile element in plant is ____________________.**

a) Calcium
b) Potash
c) nitrogen

221. **In ______________ nutrient content is expressed in ratios of nutrients.**
   a) DRIS
   b) STCR
   c) IPNM

222. **In Gujarat, salts affected soils observed in ______ region is due to marine deposit of receding of the sea.**
   a) Bhal
   b) Charotar
   c) Ghed
   d) Kachchh

223. **In waterlogged rice field, ___ are capable of fixing atmospheric nitrogen to the soil.**
   a) Fungi
   b) Blue-green algae
   c) Bacteria
   d) Actenomycetes

224. **Irrigation water having SAR value 10-18 is considered as ________________ Na water.**
   a) Low
   b) Medium
   c) High
   d) Very high

225. **Kaolinite and illite types of clay minerals are dominant in _______soils.**
   a) Acid soils
   b) Alkali soils
   c) Saline soils
   d) Saline-sodic

226. **Kaolinite is ____________ type of clay mineral.**
   a) Expanding
   b) Non-expanding
   c) Less expanding
   d) Vone of these

227. **kg/ha = ppm x ___________**
   a) 0.864
   b) 2.24
   c) 1.20
   d) 2.0

228. **Khaira disease of rice is caused by ____________deficiency.**
   a) Zinc
   b) Iron
   c) Boron
   d) Nitrogen

229. **Kjeldahl method is used to determine __________ of soil.**
   a) Total N b) Organic C
   c) Available K d) Available P

230. **Lime induced chlorosis is mostly observed in __________,**
   a) Acid soils b) Alkali soils
   c) Calcareuses soils d) Saline

231. **Lime requirement can be determined by__________ method.**
   a) Olsen's B) Schoonover's
   c) Walkly & Black d) Parker's

232. **Mahua cake contain __________ alkaloid.**
   a) Nimbidin b) Ricin
   c) Saponin d) Non of these

233. **mg/100 g = me/100 g x __________ .**
   a) Mol. wt. b) eq. wt
   c) Atomic wt. d) Atomic no.

234. **Most of the gaseous interchange in soil occurs by ________.**
   a) Mass flow b) Diffusion
   c) Interception d) Contact exchange

235. **Movement of nutrient ions and salts along with moving water is termed as:**
   a) Diffusion b) Contact exchange
   c) Mass flow d) Contact exchange

236. **Natural aggregates are called as __________.**
   a) Clods b) Peds
   c) Sphere d) Pebbles

237. **One tone of gypsum is equivalent to ________ tone of sulphur.**
   a) 0.570 b) 0.186
   c) 1.620 d) 2.300

238. **Only _______ per cent of mineral nutrients are dissolved in soil water.**

a) < 0.2 b) < 2.0

c) > 2.0 d) <20.0

239. **Total porosity is more in______ .**

a) Sandy soil b) Loamy Soil

c) Clayey soil d) Sandy loam

240. **Out of A, B and C horizon, _______ is not a part of soil solum.**

a) B b) C

c) A d) All above

241. **Parts per million (ppm)=_______**

a) mg/kg b) g/kg

c) mg/g d) kg/kg

242. **Plant samples are oven dried at ________ °C.**

a) 105 + 5 b) 85 + 5

c) 65 + 5 d) 70+ 5

243. **Red colour of soil is associated with _________.**

a) Ferric oxide b) Silica

c) Lime d) Feldspars

244. **Relative proportion of sand, silt and clay is referred as _________.**

a) Soil texture b) Soil structure

c) Soil pores d) Plasticity

245. **Rhizobium fixes _________ kg atmospheric nitrogen /ha to the soil.**

a) 10 b) 20

c) 30 d) 40

246. **Rock phosphate is applied in ____________ type soil.**

a) Alkali b) Acid

c) neutral d) Saline-alkali

247. **Salt moved to inland through the transportation of spray by wind are called ________ .**

a) Cyclic salt b) Salt deposits
c) Transported salts d) Secondary salt

248. **Soil act as a __________ filter by adsorbing and precipitation.**

a) Physical b) Biological
c) Chemical d) Physico- chemical

249. **Soil pH > 8.5 indicates soil is_______**

a) Alkaline b) Acidic
c) Neutral d) None

250. **Soils having EC - rated as Saline soil**

a) EC > 4.0 b) EC < 4.0
c) EC 1.0

251. **Spectrophotometer is used for the determination of ___________.**

a) Available P b) Available K
c) Available N d) Available S

252. **Stokes' law is applied in estimation of______.**

a) Electrical conductivity b) Mechanical analysis of soil
c) Soil pH d) EC

253. **Sulphate of potash contain ___________ % K_2O.**

a) 40-44 % b) 48-52 %
c) 52-56 % d) 36-40 % (60-62%

254. **The __________ is example of non-metal nutrient.**

a) Mg b) Mn
c) Mo d) Zn

255. **The active acidity of soils is measured by __________**

a) EC b) pH
c) ESP d) RSC

256. The activities of ________ more in acid soils.

a) Bacteria b) Actinomycetes

c) Fungi d) Algae

257. The application of fertilizer in irrigationwater in either open or closed system is known as ________.

a) Fertigation b) Drip irrigation

c) Sprinkler irrigation d) Flood irrigation

258. The arrangement of soil particle into certain definite pattern is called:

a) Soil texture b) Soil structure

c) Soil consistence d) Plasticity

259. The arrangement of three atoms in a water molecule is not ______.

a) Symmetrical b) Asymmetrical

c) Equal d) double

260. The author of "Introductory Soil Science" is ________.

a) N. C. Brady b) D. K. Das

c) J. S. Kanwar d) T.D. Biswas

261. The author of "Soil Fertility -Theory and Practice" is ________.

a) Gupta b) Kanwar

c) Tisdale d) T.D. Biswas

262. The bulk density of organic matter is about ________ g/cc.

a) 1.0 b) **1.5**

c) 0.5 d) 2.0

263. The bulk density of soil is always ___ than particle density in cultivated soil.

a) Less b) Equal

c) Greater d) Triple

264. The CEC of illite is ranged between ________ me/100g.

a) 15-40 b) 3-15
c) 80-100 d) 120-150

265. The chelated form of zinc is__________.

a) Zn-EDTA b) Zn-DTPA
c) Zn-ZnO d) Zn-CDTA

266. The chemical formula of gypsum is ________

a) $CaCl_2.2H_2O$ b) $CaSO_4.2H_2O$
c) $CaCO_3$ d) $CaSO_4$

267. The chemical formula of lime is ________

a) $CaCl_2.2H_2O$ b) $CaSO_4.2H_2O$
c) $CaCO_3$ d) $CaSO_4$

268. The clay colloids can be moulded in any shape due to the property of_____ .

a) Consistency b) Swelling
c) Plasticity d) Shrinkage

269. The clay minerals carry __________ electrical charge on their surface.

a) Positive b) Negative
c) Neutral d) None

270. The CO_2 content of the cultivated soil is _______ per cent.

a) 0.25 b) 0.03
c) 20.60 d) 30.0

271. The colloidal particles are coagulated by adding an oppositely charged ion and formation of flocs is known as __________.

a) Flocculation b) Deflocculation
c) Adsorption. d) Absorption.

272. **The criteria of nutrient essentiality are given by ________.**
a) Schoonover b) Arnon
c) Jackson d) Schofield

273. **The dicot plant roots have ____________ CEC.**
a) High b) Medium
c) Low d) None of these

274. **The excavation and destructive action of water on rocks and minerals is called___.**
a) Exfoliation b) Denudation
c) Chemical weathering d) physical

275. **The horizons A + B + C collectively known as_____ .**
a) Solum b) Regolith
c) Parent material d) Topography

276. **The irrigation water having class _____ can be use for most soils and most crops.**
a) C_2S_2 b) C_1S_1
c) C_4S_4 d) C_3S_4

277. **The lightness or darkness of soil colour is denoted by_____.**
a) Hue b) Value
c) Chroma d) All above

278. **The metamorphism of shale to slate is due to_____.**
a) Pressure b) Heat
c) Water d) Time

279. **The minute colloidal clay particle is technically called _____.**
a) Micelle b) Mineral
c) Atom d) Atomic

280. **The most abundant mineral in soil is______**
a) Quartz b) Feldspar
c) Silicates d) Alimonium

281. **The movement of water through column of soil is known as_____.**
   a) Infiltration b) Hydraulic Conductivity
   c) Percolation

282. **The native element of diamond is __________.**
   a) O b) H
   c) C d) N

283. **The nitrogen content in FYM generally ranged from _________ %.**
   a) 0.05 - 10 b) 5 - 10
   c) 0.5 - 1.0 d) 10-15

284. **The nitrogen content in Urea is _________ per cent.**
   a) 24 b) 46
   c) 64 d) 48

285. **The optimum range of temperature for the activity of micro-organisms is___.**
   a) 20-25° C b) 25-30° C
   c) 30-35° C d) 10-15° C

286. **The parent materials of sand size transported by the action of wind are called_____.**
   a) Eolian b) Moraine
   c) Loess d) Colostrums

287. **The pH is defined as the negative logarithm of _____________activity.**
   a) Nitrogen b) Calcium
   c) Hydrogen d) Nitrogen

288. **The physical weathering is also known as ________.**
   a) Decomposition b) Integration
   c) Disintegration d) All above

289. **The process of accumulation of soluble salt in soil is known as______.**
   a) Alkalization b) Solonisation
   c) Salinization d) Solodization

290. **The process of determining the amount of sand, silt and clay is known as _______ analysis.**
   a) Soil b) Mechanical
   c) Statistical d) Chemical

291. **The residual effect of urea on soil reaction is___________.**
   a) Neutral b) Acidic
   c) Basic d) None of these

292. **The saline - alkali soil is also known as ___________.**
   a) White salt b) Cyclic salt
   c) Usar d) Acidic

293. **The size of clay particle of the soil _______ mm.**
   a) < 0.02 b) < 0.002
   c) < 0.2 d) 2.00

294. **The sodium adsorption ratio (SAR) developed by _________.**
   a) USSR b) USA
   c) USSSL

295. **The soil having PD 2.50 g/cc and BD 1.25 g/cc will have_____ % porosity.**
   a) 45 b) 50
   c) 55

296. **The soil order _______ contains organic matter more than 30 %.**
   a) Aridisols b) Histosols
   c) Vertisols d) Aridisol

297. **The solid zone of the earth's sphere is known as ________.**

a) Atmosphere b) Hydrosphere
c) Lithosphere d) Stratosphere

298. **The source of N - fertilizer recommended for Tobacco**

a) Urea b) Ammonium Sulphate
c) Potassium Nitrate d) CAN

299. **The specific gravity of light mineral is less than_____.**

a) 2.65 b) 2.85
c) 2.55 d) 2.65

301. **The toxic ingredient present in urea is ____________.**

a) Uret b) Biuret
c) Themate d) Acid

302. **The value of soil EC indicates _________ in soil.**

a) H ion b) Insoluble salts
c) Soluble salts d) Al ion

303. **The vertical cracks are major problem in_______.**

a) Loamy soils b) Sandy soils
c) Deep black soils d) Silt

304. **The water held tightly on the soil surface colloidal particle is known as____.**

a) Available water b) Capillary water
c) Hygroscopic water d) Field capacity

305. **The zeta potential is the magnitude of ________charge on the colloidal particles.**

a) Neutral b) Positive
c) Negative d) None

306. **The zinc content in zinc sulphate is __________ per cent.**

a) 11 b) 21
c) 31 d) 41

307. **There are _________ categories of classification in soil taxonomy.**

a) Three b) Five
c) Six d) Eightr

308. **Thompson and Way (1982) first recognized _______ exchange phenomenon.**

a) Cation b) Anion
c) Acid d) Base

309. **To categorize the soil into high available P_2O_5 , the value should be more than __________ kg/ha.**

a) 28 b) 56
c) 42 d) 62

310. **Total porosity is more in______ .**

a) Sandy soil b) Loamy Soil
c) Clayey soil d) Sandy loam

311. **Two water molecules are held by __________bond.**

a) Hydrogen b) Ionic
c) Covalent d) None

312. **Water plays an important role in the formation of______.**

a) Igneous rock b) Sedimentary rock
c) Metamorphic rock d) All above

313. **When the magma solidified at greater depth is called_________.**

a) Effusive b) Intrusive
c) Plutonic d) None

Answers

1	c	2	c	3	c	4	c	5	d	6	b	7	a	8	b	9	b	10	b
11	a	12	a	13	a	14	c	15	a	16	a	17	b	18	c	19	a	20	c
21	a	22	a	23	a	24	b	25	a	26	b	27	a	28	b	29	a	30	a
31	a	32	b	33	c	34	a	35	b	36	a	37	c	38	d	39	a	40	c
41	b	42	b	43	c	44	c	45	b	46	a	47	c	48	c	49	b	50	b

51 b	52 c	53 b	54 b	55 a	56 c	57 c	58 c	59 b	60 b
61 a	62 b	63 a	64 b	65 c	66 b	67 a	68 b	69 b	70 a
71 c	72 c	73 c	74 c	75 d	76 c	77 b	78 c	79 c	80 c
81 b	82 c	83 c	84 b	85 a	86 c	87 a	88 b	89 c	90 c
91 a	92 c	93 b	94 c	95 a	96 b	97 b	98 a	99 b	100 b
101 b	102 a	103 b	104 b	105 c	106 b	107 a	108 b	109 c	110 b
111 c	112 c	113 a	114 b	115 c	116 b	117 c	118 a	119 a	120 c
121 a	122 b	123 c	124 a	125 c	126 b	127 b	128 c	129 b	130 b
131 a	132 d	133 a	134 c	135 b	136 b	137 b	138 c	139 a	140 a
141 c	142 a	143 b	144 a	145 b	146 b	147 a	148 c	149 d	150 b
151 a	152 a	153 b	154 b	155 c	156 c	157 b	158 c	159 c	160 c
161 b	162 a	163 b	164 b	165 a	166 a	167 a	168 a	169 b	170 a
171 c	172 a	173 c	174 c	175 c	176 b	177 c	178 b	179 a	180 b
181 a	182 a	183 a	184 a	185 a	186 a	187 a	188 c	189 b	190 a
191 a	192 c	193 b	194 a	195 c	196 c	197 c	198 b	199 a	200 b
201 c	202 c	203 c	204 c	205 c	206 b	207 c	208 a	209 b	210 a
211 b	212 b	213 c	214 c	215 a	216 b	217 c	218 b	219 c	220 a
221 a	222 a	223 b	224 b	225 a	226 b	227 b	228 a	229 a	230 c
231 b	232 c	233 b	234 b	235 c	236 b	237 b	238 a	239 c	240 c
241 c	242 c	243 a	244 a	245 d	246 b	247 a	248 c	249 a	250 a
251 a	252 b	253 b	254 c	255 b	256 c	257 a	258 b	259 b	260 b
261 b	262 c	263 a	264 a	265 a	266 b	267 c	268 c	269 b	270 a
271 a	272 b	273 a	274 b	275 a	276 b	277 b	278 a	279 a	280 c
281 c	282 a	283 c	284 b	285 b	286 a	287 c	288 c	289 c	290 b
291 b	292 c	293 b	294 c	295 b	296 b	297 c	298 c	299 b	300 b
301 b	302 c	303 c	304 c	305 c	306 b	307 c	308 a	309 b	310 c
311 a	312 b	313 b							

❑❑❑

Chapter 16

Statistics

1. **Mean is a measure of: location (central value)**
   a) Dispersion
   b) Correlation
   c) Regration
   d) None of the above

2. **Which of the followings is a measure of central value?**
   a) Median
   b) Standard deviation
   c) Mean deviation
   d) Quartile deviation

3. **Which of the followings represents median?**
   a) First quartile
   b) Fiftieth percentile
   c) Sixth decile
   d) None of the above

4. **If a constant value 50 is subtracted from each observation of a set, the mean of the series is:**
   a) Increased by 50
   b) Decreased by 50
   c) Is not affected
   d) Zero

5. **The extreme value have no effect on :**
   a) Average
   b) Median
   c) Geometric mean
   d) Harmonic mean

6. **If the minimum value in a set is 9 and its range is 57, the maximum value of the set is**
   a) 33
   b) 66
   c) 48
   d) None of the above

7. **If each value of a series is divided by 2, the mean of new values:**
   a) Is two times the original mean
   b) Is decreased by 2

c) Is increased by 2
d) Is half of the original mean

8. **If the values of a set are measured in cms., the unit of variance will be:**
a) No unit
b) cm
c) cm^2
d) cm^3

9. **Which measure of dispersion has a different unit other than the unit of measurement of Values:**
a) Range
b) Mean deviation
c) Standard deviation
d) Variance

10. **The average of the sum of squares of the deviations about mean is called:**
a) Variance
b) Absolute deviation
c) Standard deviation
d) Mean deviation

11. **Quartile deviation is equal to:**
a) Interquartile range
b) Double the interquartile range
c) Half of the interquartile range
d) None of the above

12. **Which measure of dispersion can be calculated in case of open end intervals?**
a) Range
b) Standard deviation
c) Coefficient of variation
d) Quartile deviation

13. **Which of the following is unit less measure of dispersion ?**
a) Standard deviation
b) Mean deviation
c) Coefficient of Variation
d) Range

14. **Three houses were available in a locality for allotment. Three persons applied for a house. The probability that all the three persons applied for the same house is:**

a) 1/3 b) 1/9
c) 1/27 d) 1

15. **A speaks truth 4 times out of five and B speaks truth 3 times out of four. They agree in the assertion that a white ball has been drawn from a bag containing 10 balls of different colours. The probability that a white ball was really drawn is:**

a) 3/50 b) 1/27
c) 1/1350 d) 81/82

16. **In the problem of question 15 if the bag contains 1 white and 9 red balls, the probability of one white ball being drawn is:**

a) 4/7 b) 3/50
c) 9/200 d) None of the above

17. **If A tells truth 4 times out of 5 and B tell truth 3 times out of 4. The probability that, both expressing the same fact contradicts each other is:**

a) 1/20 b) 3/20
c) 1/5 d) None of the above

18. **The family of parametric distributions which has mean always less than variance is:**

a) Beta distribution b) Lognormal distribution
c) Weibull distribution d) Negative binomial distribution

19. **The distribution possessing the memory less property is:**

a) Gamma distribution
b) Geometric distribution
c) Hypergeometric distribution
d) All the above

20. **Which one problem out of the four is not related to stratified sampling?**

a) Fixing the criterion for stratification
b) Fixing the number of strata

c) Fixing the sample size
d) Fixing the points of demarcation between strata

21. **Regarding the number of strata, which statement is true?**
a) Less the number of strata, better it is
b) More the number of strata, poorer it is
c) More the number of strata, better it is
d) Not more than ten items should be there in a stratum

22. **Under equal allocation in stratified sampling, the sample from each stratum is:**
a) Proportional to stratum size
b) Of same size from each stratum
c) In proportion to the per unit cost of survey of the stratum
d) All the above

23. **Under proportional allocation, the size of the sample from each stratum depends on:**
a) Total sample size
b) Size of the stratum
c) Population size
d) All the above

24. **Under proportional allocation one get:**
a) An optimum sample
b) A self-weighing sample
c) Both a) and b)
d) Neither a) nor b)

25. **How many types of optimum allocation are in common use?**
a) One
b) Two
c) Three
d) Four

26. **Least square estimators of the parameters of a linear model are:**
a) Unbiased
b) BLUE
c) UMVU
d) **All the above**

27. Least square estimators of the parameters of a linear model are not:

a) Necessarily consistent b) Scale invariant

c) Asymptotically normal d) All the above

28. A hypothesis may be classified as:

a) Simple b) Composite

c) Null d) All the above

29. The hypothesis under test is:

a) Simple hypothesis b) Alternative hypothesis

c) Null hypothesis d) None of the above

30. Whether a test is one sided or two sided depends on:

a) Alternative hypothesis b) Composite hypothesis

c) Null hypothesis d) Simple hypothesis

31. A wrong decision about Ho leads to:

a) One kind of error b) Two kinds of error

c) Three kinds of error d) Four kinds of error

32. Power of a test is related to:

a) Type I error

b) Type II error

c) Types I and II errors both

d) None of the above

33. If q is the true parameter and b the type II error, the function b(q) is known as:

a) Power function

b) Power of the test

c) Operating characteristic function

d) None of the above

34. In terms of type II error b and q, the true parameter, the function 1 -b (q) is called:

a) Power of the test
b) Power function
c) OC function
d) None of the above

35.Out of the two types of error in testing, the more severe error is:

a) Type I error
b) Type II error
c) Both **a) and b) are equally severe-**
d) No error is severe

36. Area of the critical region depends on:

a) Size of type I error
b) Size of type II error
c) Value of the statistic
d) Number of observations

37. Regression coefficient is independent of

a) Origin
b) Scale
c) Both a) & b)
d) Nither a) nor b)

38. Randomization in an experiment helps to eliminate:

a) Systematic influences
b) Human biases
c) Dependence among observations
d) All the above

39. Binomial and Poisson distribution are of :

a) Discrete type
b) Continuous type
c) Discrete as well as continuous
d) None of above

40. Normal distribution curve with regard to bulginess :

a) leptokurtic
b) mesokurtic
c) platykurtic
d) not definite

41. **The relation between the mean and variance of chi-square dist^n with n d.f. is**

a) Mean = 2 variance b) 2 mean = variance

c) Mean = variance d) None of the above

42. **Most of the non-parametric methods utilize measurements on**

a) Interval scale b) Ratio scale

c) Ordinal scale d) Nominal scale

43. **Two random variables X and Y are said to be independent if**

a) E(XY) = 1 b) E(XY) = 0

c) E(XY) = E(X).E(Y) d) E(XY) = any constant value

44. **Chi-square distribution curve in respect of symmetry is**

a) Positively skew b) Negatively skew

c) Symmetrical d) Any of the above

45. **Coefficient of skewness for normal distribution is**

a) Greater than 3 b) Less than 3

c) Equal to 3 **d) Zero**

46. **If b_{yx} and b_{xy} are two regression coefficients, they have :**

a) same sign

b) opposite sign

c) either same or opposite signs

d) nothing can be said

47. **A sample consists of :**

a) all units of the population

b) 50% units of the population

c) 5% units of the population

d) any fraction of the population

48. **A population consisting of all the items which are physically present is called :**

a) Hypothetical population

b) Real population

c) Infinite population
d) None of the above

49. **A function of variates for estimating a parameter is called :**
a) An estimate
b) An estimator
c) A frame
d) A statistic

50. **If all the observations in a set of observations are same, the variance of the set of values is :**
a) Zero
b) One
c) Infinity
d) Not possible to calculate

51. **The number of all possible samples of size two from a population of 4 units as:**
a) 2
b) 4
c) 8
d) **12**

Answers

1 a	2 a	3 b	4 b	5 b	6 b	7 d	8 c	9 d	10 a
11 c	12 d	13 c	14 b	15 d	16 a	17 d	18 d	19 b	20 c
21 c	22 b	23 d	24 b	25 c	26 d	27 c	28 d	29 c	30 a
31 b	32 b	33 c	34 b	35 b	36 a	37 a	38 b	39 a	40 b
41 c	42 c	43 c	44 a	45 d	46 a	47 d	48 b	49 b	50 a
51 d									

❑❑❑